D
UNIVE...
2423 NORMAN HALL
GAINESVILLE, FL 32611

Aesthetic Challenge in Science Education

The Assessment Challenge in Statistics Education

Edited by

I. Gal
University of Haifa, Israel

and

J.B. Garfield
University of Minnesota, USA

Technical Editor:
Matthew Segaard

Amsterdam • Berlin • Oxford • Tokyo • Washington, DC

and

The International Statistical Institute, The Netherlands

© 1997, The International Statistical Institute

All rights reserved. No part of this book may be reproduced, stored in a retrieval system, or transmitted, in any form or by any means, without the prior written permission from the publisher.

ISBN 90 5199 333 1 (IOS Press)
ISBN 4 274 90158 0 C3000 (Ohmsha)
Library of Congress Control Number: 97-71649

Publisher
IOS Press
Nieuwe Hemweg 6B
1013 BG Amsterdam
The Netherlands
fax: +31 20 6870019
e-mail: order@iospress.nl

Distributor in the UK and Ireland
IOS Press/Lavis Marketing
73 Lime Walk
Headington
Oxford OX3 7AD
England
fax: +44 1865 750079

Distributor in the USA and Canada
IOS Press, Inc.
4502 Rachael Manor Drive
Fairfax, VA 22032
USA
fax: +1 703 323 3668
e-mail: iosbooks@iospress.com

LEGAL NOTICE
The publisher is not responsible for the use which might be made of the following information.

Contents

List of Contributors   vii

Preface   xi

1. Curricular Goals and Assessment Challenges in Statistics Education, *Iddo Gal and Joan Garfield*   1

Part I. Curricular Goals and Assessment Frameworks

2. Some Emerging Influences Underpinning Assessment in Statistics, *Andy Begg*   17

3. Authentic Assessment Models for Statistics Education, *Shirley Colvin and Kenneth E. Vos*   27

4. Monitoring Attitudes and Beliefs in Statistics Education, *Iddo Gal, Lynda Ginsburg and Candace Schau*   37

Part II. Assessing Conceptual Understanding of Statistical Ideas

5. A Framework for Assessing Knowledge and Learning in Statistics (K-8), *Susan N. Friel, George W. Bright, Dargan Frierson and Gary D. Kader*   55

6. Using "Real-Life" Problems to Prompt Students to Construct Conceptual Models for Statistical Reasoning, *Richard Lesh, Miriam Amit and Roberta Y. Schorr*   65

7. Simple Approaches to Assessing Underlying Understanding of Statistical Concepts, *Anthony E. Kelly, Finbarr Sloane and Andrea Whittaker*   85

8. Assessing Students' Connected Understanding of Statistical Relationships, *Candace Schau and Nancy Mattern*   91

Part III. Innovative Models for Classroom Assessment

9. Assessing Statistical Thinking Using the Media, *Jane M. Watson* — 107

10. Assessing Students' Statistical Problem-Solving Behaviors in a Small-Group Setting, *Frances R. Curcio and Alice F. Artzt* — 123

11. Assessing Students Projects, *Susan Starkings* — 139

12. Assessing Project Work by External Examiners, *Peter Holmes* — 153

13. Portfolio Assessment in Graduate Level Statistics Courses, *Carolyn M. Keeler* — 165

14. Technologies for Assessing and Extending Statistical Learning, *Susanne P. Lajoie* — 179

15. Issues in Constructing Assessment Instruments for the Classroom, *Flavia Jolliffe* — 191

16. Assessment on a Budget: Using Traditional Methods Imaginatively, *Chris Wild, Chris Triggs and Maxine Pfannkuch* — 205

Part IV. Assessing Understanding of Probability

17. Dimensions in the Assessment of Students' Understanding and Application of Chance, *Kathleen E. Metz* — 223

18. Combinatorial Reasoning and its Assessment, *Carmen Batanero, Juan D. Godino and Virginia Navarro-Pelayo* — 239

19. Probability Distributions, Assessment and Instructional Software: Lessons Learned from an Evaluation of Curricular Software, *Steve Cohen and Richard A. Chechile* — 253

References — 263

Author Index — 277

Subject Index — 281

List of Contributors

Miriam Amit: Mathematics Supervision, Ministry of Education, Israel (amit@bgumail.bgu.ac.il)

Alice F. Artzt: School of Education, Queens College of the City University of New York, USA (alaqc@qcvaxa.acc.qc.edu)

Carmen Batanero: Department of Mathematics Education, University of Granada, Spain (batanero@goliat.ugr.es)

Andy Begg: Centre for Science, Mathematics and Technology Education Research, University of Waikato, New Zealand (a.begg@waikato.ac.nz)

George W. Bright: School of Education, University of North Carolina at Greensboro, USA (brightg@steffi.uncg.edu)

Richard A. Chechile: Psychology Department, Tufts University, USA (rchechil@emerald.tufts.edu)

Steve Cohen: Curricular Software Studio, Tufts University, USA (scohen@emerald.tufts.edu)

Shirley Colvin: Columbia Elementary School, Columbia, Maryland, USA

Frances R. Curcio: School of Education, New York University, USA (curcio@is2.nyu.edu)

Susan N. Friel: School of Education, University of North Carolina at Chapel Hill, USA (sfriel@email.unc.edu)

Dargan Frierson: University of North Carolina at Wilmington, USA (frierson@uncwil.edu)

Iddo Gal: Department of Human Services, University of Haifa, Israel (iddo@research.haifa.ac.il)

Joan Garfield: Department of Educational Psychology, University of Minnesota, USA (jbg@maroon.tc.umn.edu)

Lynda Ginsburg: Graduate School of Education and National Center on Adult Literacy, University of Pennsylvania, USA (ginsburg@literacy.upenn.edu)

Juan D. Godino: Department of Mathematics Education, University of Granada, Spain (jgodino@goliat.ugr.es)

Peter Holmes: Royal Statistical Society Centre for Statistical Education, University of Nottingham, United Kingdom (ph@pmn1.maths.nott.ac.uk)

Flavia Jolliffe: Department of Computing and Mathematical Sciences, University of Greenwich, United Kingdom (f.r.jolliffe@greenwich.ac.uk)

Gary D. Kader: Department of Mathematical Sciences, Appalachian State University, USA (kadergd@conrad.appstate.edu)

Carolyn M. Keeler: Department of Educational Administration, University of Idaho at Boise, USA (ckeeler@uidaho.edu)

Anthony E. Kelly: Graduate School of Education, Rutgers University, USA (aekelly@rci.rutgers.edu)

Susanne P. Lajoie: Department of Educational and Counselling Psychology, McGill University, Canada (lajoie@education.mcgill.ca)

Richard Lesh: Department of Mathematics, University of Massachusetts at Dartmouth, USA (rlesh@umassd.edu)

Nancy Mattern: College of Education, University of New Mexico, USA (nmat@unm.edu)

Kathleen E. Metz: School of Education, University of California at Riverside, USA

Virginia Navarro-Pelayo: Department of Mathematics Education, University of Granada, Spain

Maxine Pfannkuch: Department of Mathematics and Department of Statistics, University of Auckland, New Zealand (pfannkuc@mat.auckland.ac.nz)

Candace Schau: College of Education, University of New Mexico, USA (cschau@unm.edu)

Roberta Y. Schorr: Center for Mathematics, Science, and Computer Education, Rutgers University, USA (schorr@rci.rutgers.edu)

Finbarr Sloane: Hewitt Associates, Chicago, USA (fcsloane@hewitt.com)

Susan Starkings: School of Computing, Information Systems and Mathematics, South Bank University, United Kingdom (starkisa@vax.sbu.ac.uk)

Chris Triggs: Department of Statistics, University of Auckland, New Zealand (triggs@stat.auckland.ac.nz)

Kenneth E. Vos: Department of Education, College of St. Catherine, St. Paul, USA
(kevos@alex.stkate.edu)

Jane M. Watson: School of Education, University of Tasmania, Australia
(Jane.Watson@educ.utas.edu.au)

Andrea Whittaker: Psychologist, San Jose, California, USA
(akwhitt@aol.com)

Christopher J. Wild: Department of Statistics, University of Auckland, New Zealand
(wild@stat.auckland.ac.nz)

Preface

This book discusses conceptual and pragmatic issues in the assessment of statistical knowledge, reasoning skills, and dispositions of students in diverse contexts of instruction, both at the college and precollege levels. It is designed primarily for academic audiences interested in the teaching and learning of statistics and mathematics and for those involved in teacher education and training in diverse contexts.

This project was initiated because we sensed that assessment is a critical issue that has received little explicit attention in the statistics education literature. In developing this volume, we solicited chapters from a variety of authors with expertise in statistics education, mathematics education, assessment, educational research, teacher training, and curriculum development. We sought colleagues able to address both theoretical and applied aspects of teaching and assessing students' cognitions and achievement. We asked our authors to address classroom realities and consider the challenges involved in implementing new methods with students and teachers who study statistics in academic or K-12 contexts.

To make the most out of this book, the following points may be useful:

- Chapter 1 may be a good starting point. It outlines common curricular goals in statistics education that underlie the discussions in all chapters in this volume and contains an overview of the remaining chapters.

- The editorial process ensured that all chapters open with a brief *Purpose* statement that lists the key questions the chapter addresses, and end with an *Implications* section that reiterates key ideas and challenges raised. Readers can examine the *Purpose* and *Implications* sections for a quick determination of the chapter's contents.

- Several chapters address assessment in the context of a specific age or student group (e.g., elementary, secondary, graduate), and examples of tasks and student responses are described in a specific context of instruction. Yet, statistics is a unique subject in that many of the basic topics (e.g., displaying and making sense of data, using summary statistics, or drawing inferences), are presented in much the same way regardless of the level of instruction. Authors were asked to make sure their points are of relevance to a broad range of readers and to discuss adaptations for different age groups where relevant. Therefore, we believe that readers interested in one age group or level of instruction may find much relevant material in other chapters that address a different student population or teaching context.

- Given the international audience for this book, the editorial process attempted to remove country-specific terminology while preserving the authenticity and uniqueness of chapters couched in different cultural contexts. As a result, readers may perceive the use of different terms that essentially mean the same thing (e.g., scoring, grading, or marking).

A project of this magnitude would have been impossible without the help of numerous individuals and organizations.

Many thanks go to the National Center on Adult Literacy (NCAL), a federally-supported R&D center at the Graduate School of Education of the University of Pennsylvania, and its director, Daniel Wagner, for supporting Iddo's work on numeracy and statistical issues until his move in 1995 to Israel, and for generously hosting two working conferences in 1992 and 1994 on statistics assessment. These meetings, which we co-directed, informed our work as well as the writings of some of those who contributed to this volume.

The National Science Foundation (NSF) funded parts of our work in the area of statistics since 1990, as well as the 1992 and 1994 working meetings, through grants MDR90-50006 and RED-9355860. This funding, part of NSF's long-standing investment in promoting mathematics and science education, has been pivotal in enabling us to sustain our effort through the three years it took to complete this project. (Of course, NSF is not responsibile for any ideas and opinions presented here.)

We want to acknowledge the financial help received from the National Center for Research in Mathematical Sciences Education (NCRMSE) at the University of Wisconsin-Madison, and the encouragement provided by its director, Tom Romberg, and by Susanne Lajoie, who served as the co-director of NCRMSE's statistics working group.

The International Statistical Institute (ISI) and its Director, Zoltan Kennessy, as well as the International Association on Statistics Education and its President, Anne Hawkins, have been instrumental in providing a publishing venue for this book. Daniel Berze, Director of Publications for ISI, has skillfully managed the publication on behalf of ISI and IOS Press. The University of Pennsylvania, the University of Haifa and the University of Minnesota have supported our work and the inordinate amount of copying, faxing and e-mail traffic required.

Most importantly, we would like to thank our many contributors, who remained focused on the goal of sharing their experiences and insights with the educational community, while enduring multiple review cycles and editing demands needed to fit their contributions into a limited space.

Our gratitude and appreciation go to Matthew Segaard, our technical editor at the University of Minnesota, who under a tight production schedule diligently and ably worked with Joan Garfield on the many large and small editing chores required to prepare a finished camera-ready copy.

Lastly, to Aviva Gal and to Michael Luxenberg, our spouses, and to our children Naama and Danna, and Rebecca and Harlan, our many thanks for bearing with the incessant computer clutter and with the many hours it took to complete this project.

The 19 chapters in this volume by no means exhaust all assessment issues encountered in environments where statistics is being taught. Yet, taken as a whole, the chapters constitute a rich resource that for the first time offers, beyond a conceptual basis for teaching statistics, solid rationales and a wealth of suggestions for assessing students' achievement and progress regarding knowledge, skills, reasoning, and dispositions. We hope that this volume will stimulate a scholarly discourse within the professional community involved in statistics education, and that in coming years additional publications will examine in more detail the many assessment and teaching needs and challenges raised in this volume.

Iddo Gal
Joan Garfield

Chapter 1

Curricular Goals and Assessment Challenges in Statistics Education

Iddo Gal and Joan Garfield

Purpose

This chapter frames the main issues with which this volume deals. The chapter examines common goals for statistics education at both precollege (school) and college levels, describes the resulting challenges for assessment in statistics education, and outlines the main issues addressed by each of the chapters in this volume. Finally, needs for future research and development are discussed.

INTRODUCTION

Statistics has gained recognition as an important component of the precollege mathematics and science curriculum. Establishing a place in the elementary and secondary curriculum led to the production of new instructional materials for elementary and secondary schools (e.g., Landwehr & Watkins, 1987; Travers, Stout, Swift & Sextro, 1985; Friel, Russell & Mokros, 1990; Konold, 1990; COMAP, 1990; Ohio Math Project, 1992). At the college level, where statistics courses have traditionally been taught, changes in content and pedagogy are being recommended as part of a "statistics reform" effort (e.g., Cobb, 1992; Moore, in press).

Now that attention to the teaching of statistics has become visible at all educational levels, it has become apparent that assessment of student learning and understanding of statistics is not being adequately addressed in current projects and instructional efforts. Until recently (see Garfield, 1994; Konold, 1995; Gal & Ginsburg, 1994) very few publications addressed assessment issues in statistics education. This lack of information is alarming in light of extensive research showing that statistics and probability concepts are difficult to teach and often poorly understood (Garfield & Ahlgren, 1988; Shaughnessy, 1992).

Our goal in creating this book was to provide a useful resource to educators and researchers interested in helping students at all educational levels to develop statistical knowledge and reasoning skills. We advise readers focused on students at one level (e.g., secondary) to not skip over chapters describing students at other levels. We are convinced that students who are introduced to statistical ideas and procedures learn much the same material and concepts (e.g., creating graphical displays of data, describing the center and dispersion of data, inference from data, etc.) regardless of their grade level. Therefore, we believe that discussions of assessment issues couched in the reality of one age group will be of interest to those interested in instruction with other types of students.

This introductory chapter aims to establish a common base for the remaining chapters of this volume, and is organized in three parts. First, eight instructional goals in statistics education that are common to all or most levels of instruction are discussed and some tensions inherent in attempting to reach these goals are explored. Next, several challenges in assessment of students' progress towards these goals are noted and the unique issues addressed by each chapter are outlined. Finally, implications for needed practices, development, and research are outlined.

INSTRUCTIONAL GOALS

In a book discussing assessment of students' knowledge and understanding of any subject, it is desirable to start by presenting a coherent framework of the target concepts, knowledge, or understanding that are to be learned, and therefore, to be assessed. This section aims to identify common goals that apply across diverse contexts of instruction and educational levels and that are presupposed by many of the chapters in this book.

At the precollege level, statistics may be taught to elementary or secondary students, usually as part of the mathematics curriculum. It may be taught as a course, a single unit, as part of a classroom project, or embedded in a mathematics topic such as functions or graphing. The NCTM *Curriculum and Evaluation Standards* (1989), and Project 2061's *Benchmarks for Science Literacy* (AAAS, 1993), both outline broad learning goals for statistics and probability. These and related sources leave wide latitude for interpretation by individual teachers, curriculum developers, and teacher-training programs regarding specific curricular goals and preferred assessment methods.

As a postsecondary subject, many students encounter statistics in a stand-alone generic course (i.e., the infamous "introductory statistics" service course), but some are introduced to statistics couched within the needs and applications of a specific field, such as psychology, business, economics, or engineering. Statistics also continues to be taught as part of the core curriculum in most mathematics departments. These postsecondary versions of introductory statistics courses differ quite a bit in their goals for students, the level of mathematical background required, the coverage of probability theory, and the use of technology. However, general dissatisfaction with student outcomes across courses led to a task force report, commissioned by the Mathematical Association of America (Cobb, 1992). This report offered recommendations for the "reform" of most introductory statistics courses. The basic message was "more data and concepts, less theory, fewer recipes." Statistics instructors were also urged to include more active learning opportunities and to decrease the amount of lecturing.

We believe that there are some common goals to instruction in these seemingly diverse educational levels and contexts. If we think broadly about what it is that we want our students to learn and be able to do with their knowledge, an overarching goal of statistics education emerges. This goal is that, by the time students finish their encounters with statistics, they become informed citizens who are able to:

- Comprehend and deal with uncertainty, variability, and statistical information in the world around them, and participate effectively in an information-laden society.

- Contribute to or take part in the production, interpretation, and communication of data pertaining to problems they encounter in their professional life.

This is a broadly defined instructional vision, and one whose achievement may extend over several years or levels of schooling or over several statistics courses.

As part of achieving this broad vision, eight interrelated basic subgoals for statistics instruction are described below. These subgoals were gleaned from our own prior work and writings (e.g., Garfield & Ahlgren, 1988; Garfield, 1995a; Gal, 1993; Gal & Baron, 1996) and from relevant literature on mathematics, statistics, and science education (e.g., NCTM, 1989; Moore, 1990; AAAS, 1993; SCANS, 1990; ASA-NCTM/Burrill, 1994; Cobb, 1992). To be sure, these subgoals are described as "basic" since they relate to the primary phase of students' encounter with statistics in most levels of instruction. Additional subgoals (e.g., regarding statistical inference or probability) may be posed for specific student populations or contexts of instruction, and it is not assumed that statistics educators will find these subgoals equally important for all student populations. However, all chapters in this volume address issues in assessment related to at least one of these basic subgoals.

Goal 1: Understand the purpose and logic of statistical investigations

Students should understand why statistical investigations are conducted, and the "big ideas" that underlie approaches to data-based inquiries. These ideas include:

- The existence of variation
- The need to describe populations by collecting data
- The need to reduce raw data by noting trends and main features through summaries and displays of the data
- The need to study samples instead of populations and to infer from samples to populations
- The logic behind related sampling processes
- The notion of error in measurement and inference, and the need to find ways to estimate and control errors
- The need to identify causal processes or factors
- The logic behind methods (such as experiments) for determining causal processes

Goal 2: Understand the process of statistical investigations

Students should understand the nature of and processes involved in a statistical investigation and considerations affecting the design of a plan for data collection. They should recognize how, when, and why existing statistical tools can be used to help an investigative process. They should be familiar with the specific phases of a statistical inquiry. These phases include (not necessarily in a linear order):

- Formulating a question
- Planning the study (e.g., approach and overall design, sampling, choice of measurement tools)
- Collecting and organizing data
- Displaying, exploring, and analyzing data
- Interpreting findings in light of the research questions
- Discussing conclusions and implications from the findings, and identifying issues for further study

Goal 3: Master procedural skills

Students need to master the "component skills" that may be used in the process of a statistical investigation. This mastery includes being able to organize data, compute needed indices (e.g., median, average, confidence interval), or construct and display useful tables, graphs, plots, and charts, either by hand or assisted by technology (e.g., a calculator, graphing calculator, or computer).

Goal 4: Understand mathematical relationships

Students should develop an understanding, intuitive and/or formal, of the main mathematical ideas that underlie statistical displays, procedures, or concepts. They should understand the connection between summary statistics, graphical displays, and the raw data on which they are based. For example, they should be able to explain how the mean is influenced by extreme values in a data set and what happens to the mean and median when data values are changed.

Goal 5: Understand probability and chance

Moore (in press) recommends that students need only an informal grasp of probability in order to follow the reasoning of statistical inference. This understanding would develop from experiences with chance behavior starting with devices (e.g., coins and dice) and leading to computer simulations. In this way students should gain an understanding of a few key ideas (Garfield, 1995b). These ideas include:

- Concepts and words related to chance, uncertainty, and probability appear in our everyday lives, particularly in the media.
- It is important to understand probabilistic processes in order to better understand (the likelihood of) events in the world around us, as well as information in the media.
- Probability is a measure of uncertainty.
- Developing a model and using it to simulate events is a helpful way to generate data to estimate probabilities.
- Sometimes our intuition is incorrect and can lead us to the wrong conclusion regarding probability and chance events.

Goal 6: Develop interpretive skills and statistical literacy

In carrying out a statistical investigation, students need to be able to interpret results and be aware of possible biases or limitations on the generalizations that can be drawn from data. We realize that most students are more likely to be consumers of data than researchers, and will seldom have to collect or analyze data as adults. Instead, they will need to be able to make sense of published results from studies and surveys reported in the media or in a workplace context. Therefore, students need to learn what is involved in interpreting results from a statistical investigation and to pose critical and reflective questions about arguments that refer to summary statistics or to data reported in the media or in project reports from their classroom peers (e.g., How reliable are the measurements used? How representative was the sample? Are the claims being made sensible in light of the data and sample?).

Goal 7: Develop ability to communicate statistically

Strong writing and speaking skills are needed if students are to effectively communicate about statistical investigations and probabilistic phenomena or processes. Good reading comprehension and communication skills are required so that students can effectively discuss or critique statistical or probabilistic arguments they encounter which claim to be based on some data (e.g., "8 out of 10 doctors use...", "there is a 20 percent chance that..."). Students should be able to use statistical and probabilistic terminology properly, convey results in a convincing way, and be able to construct proper arguments based on data or observations. They should also be able to argue thoughtfully about the validity of other people's interpretations of data or graphical displays, and raise questions about acceptability of generalizations made on the basis of a single study or a small sample.

Goal 8: Develop useful statistical dispositions

Students should develop an appreciation for the role of chance and randomness in the world and for statistical methods and planned experiments as useful scientific tools and as powerful means for making personal, social, and business-related decisions in the face of uncertainty. They should realize that the process of statistical inquiry can often lead to better conclusions than relying on anecdotal data or on their own subjective experiences or intuitions, but that this is not guaranteed. Students should also learn to adopt a questioning stance when they are faced with an argument that purports to be based on data (e.g., "all people are...") or a report of results or conclusions from a statistical investigation, survey, or empirical research.

CHALLENGES IN STATISTICS EDUCATION

The unique challenges that statistics teachers face stem from the existence of multiple subgoals as listed above, which require teachers to address a wide range of conceptually distinct issues during instruction. Educators are further challenged by the need to make sure that students understand the real-world problems that motivate statistical work and investigations, and by the need to help students become familiar with the many nuances, considerations, and decisions involved in generating, describing, analyzing, and interpreting data and in reporting findings.

Doing statistics versus being informed consumers of statistics

The eight subgoals listed above are made up of two overlapping but separate clusters. The first cluster includes the first six subgoals that deal mainly with "doing" statistics (understanding of purposes and uses, and of the logic of various procedures). A second, overlapping cluster includes the last three subgoals which are concerned with sense-making and communicative skills as well as with reflection and questioning. As Gal and Brayer-Ebby (in preparation) point out, these different clusters of subgoals "pull" educators in somewhat different directions as each has a different set of instructional implications, and teachers and textbooks are often more concerned with selected aspects of the first cluster than with the second cluster.

Statistics versus mathematics

The eight subgoals listed above emphasize a shift from traditional views of teaching statistics as a mathematical topic (with an emphasis on computations, formulas, and procedures) to the current view that distinguishes between mathematics and statistics as separate disciplines. As Moore (1992) argues, statistics is a mathematical science but is not a branch of mathematics, and has clearly emerged as a discipline in its own right, with characteristic modes of thinking that are more fundamental than either specific methods or mathematical theory.

The following points explicate some of the key differences between the two disciplines:

1. In statistics, the context motivates procedures and is the source of meaning and basis for interpretation of results of such activities. Moore (1990) points out that data should be viewed as numbers with a context.

2. The indeterminacy, "messiness," or context-boundedness of statistics is markedly different from the more precise, finite nature characterizing traditional learning in other mathematical domains.

3. Mathematical concepts and procedures are used as part of the attempt to manage or "solve" statistical problems, and some technical facility with them may be expected in certain courses and educational levels. However, the need for accurate application of computations or execution of procedures is rapidly being replaced by the need for selective, thoughtful, and accurate use of technological devices and increasingly more sophisticated software programs.

4. The fundamental nature of many (but not all) statistical problems is that they do not have a single mathematical solution. Rather, realistic statistical problems usually start with a question and culminate with the presentation of an opinion that may have different degrees of reasonableness.

5. A primary goal of statistics education is to enable students to be able to render reasoned descriptions, judgments, inferences and opinions about data, or argue about the interpretation of data, using various mathematical tools only to the degree needed. Judgments and inferences expected of students (e.g., predictions about a population based on sample data students collected in a survey they planned and conducted) very often cannot be characterized as "right" or "wrong," instead having to be evaluated in terms of quality of reasoning, adequacy of methods employed, and nature of data and evidence used, and may often depend on (or be biased by) students' world knowledge, which might be limited.

The need for alternative approaches to assessment

In light of the complex goals for students and the emerging differences between mathematics and statistics, an adequate assessment of student outcomes is not possible using only multiple choice or short answer questions. These types of questions are all too often divorced from context and focus on accuracy of statistical computations, correct application of formulas, or correctness of graphs and charts, thus assessing only one or two of the subgoals listed earlier. Questions and task formats that culminate in simple "right or wrong" answers do not reflect the nature of many statistical problems. Such tasks provide only limited information about students' statistical reasoning processes, their ability to construct or interpret statistical arguments, their understanding of the logic behind the use of certain procedures (e.g., sampling, averaging), or their ability to clearly and correctly use statistical or mathematical terminology when discussing their work or reasoning.

A range of different assessment methods is needed to provide broad information about the quality of students' thinking, communication, and reasoning processes (NCTM, 1995). Methods are needed that are appropriate not only for assessing specific skills (such as those under subgoals 3 and 4, which in part can be assessed by traditional item formats), but that also reveal students' understanding of the "big ideas" in statistics and their ability to choose and apply statistical tools appropriately when making sense of realistic data. A student's ability to correctly calculate the average of given data either manually or with the help of an electronic aid, for example, says little about her understanding of when the average is a reasonable way to summarize information, or what other statistical tools may be better suited for the task of describing the data (Gal, 1995).

The challenge faced by all educators involved in statistics education is to identify assessment methods that are able to elicit and reveal student learning corresponding to each of the eight subgoals outlined in the previous section. These methods need to gauge the degree of *integration* between students' skills, knowledge and dispositions and their ability to manage meaningful, realistic questions, problems, or situations, both as generators as well as interpreters of data, findings, or statistical messages. This challenge provides the rationale behind the creation of this book.

OVERVIEW OF THIS BOOK

Each chapter in this book discusses assessment issues pertaining to one or more of the eight subgoals, within the context of a specific educational level. However, the issues presented are usually relevant for other levels as well. Whenever possible, authors present examples for assessment tasks suitable for a range of instructional levels and analyze student responses to illustrate interpretive issues. The chapters have been grouped into four parts, each of which is summarized below.

Part I: Curricular Goals and Assessment Frameworks

The chapters in this part outline key curricular goals and desired outcomes in statistics education, describe recent changes and reforms in thinking about the desired processes of instruction in the mathematical sciences, and frame the assessment issues which all educators involved in statistics education have to address.

In Chapter 2, Begg reviews key issues in mathematics education and in education in general that underpin work on assessment in statistics education. Begg discusses the broadening of instruction in mathematics and statistics to include the key processes of problem-solving, reasoning, communicating, making connections, and using tools, introduces the emerging constructivist views on learning, and reiterates purposes and general principles for assessment of relevance to statistics education.

Colvin and Vos provide in Chapter 3 an introduction to the authentic assessment movement, which aims to measure student performance within tasks relevant to situations or problems outside of the school setting. An authentic assessment model for statistics education in primary and secondary schools is described which is also relevant for college level statistics instruction. The questions and stages involved in establishing an authentic assessment system are outlined and some dilemmas inherent in this process are noted. Examples for authentic assessment tasks and for student performance on such tasks are included to illustrate the ideas presented.

Chapter 4, Gal, Ginsburg, and Schau focus on assessment challenges that pertain to the curricular goal of developing productive dispositions and attitudes as part of statistics education. These authors argue that non-cognitive or meta-cognitive factors, such as (negative) attitudes or beliefs towards statistics can impede learning of statistics, or hinder the extent to which students will develop useful statistical intuitions that can be applied outside the classroom. The chapter examines current approaches for assessing students' attitudes and beliefs, and suggests ways to extend and integrate such assessment into ongoing instruction.

Part II: Assessing Conceptual Understanding of Statistical Ideas

Learning the many procedures, techniques, and analytic processes in statistics assumes that students understand the underlying concepts and "big ideas" (e.g., variation, visual representation of data, center, spread). However, understanding such concepts and ideas requires that students undergo a process of conceptual development, and that teachers acknowledge the presence of such a process and become familiar with its stages. The chapters in this part discuss the challenges involved in identifying learners' conceptual understanding and demonstrate assessment approaches that can be useful in this regard.

In Chapter 5, Friel, Bright, Frierson, and Kader outline a framework for thinking about what teachers and students know and should able to do with respect to learning statistics in K-8 schools. They focus on assessment of teachers' and students' understanding of concepts and ideas related to graphical representations, in light of the limited knowledge about the complexities of learning such concepts. The authors grapple with questions about the nature of "good tasks" that may be used to assess graph knowledge, and demonstrate assessment tasks that can advance our understanding of the complexities of assessing students' graph knowledge.

Lesh, Amit, and Schorr describe in Chapter 6 an approach to statistical education which focuses on models, modeling, and "model-eliciting" activities. Characteristics of realistic problems which prompt students to construct conceptual models for statistical reasoning are described. An illustrative project-size "performance assessment" activity is presented in which students work in teams, using appropriate technology-based tools, and in which students are able to simultaneously learn and document (self-assess) what they are learning. Examples for students' ways of thinking are provided, and criteria for designing model-eliciting activities that highlight students' conceptual understanding are offered.

Students' lack of conceptual understanding is the topic of Chapter 7. Kelly, Sloane, and Whittaker contend that students approach statistical terminology as a foreign language, disassociated from students' existing mathematical knowledge. This causes many students to learn statistics in a rote fashion rather than understand the underlying logic, or to adopt a series of statistical routines that are poorly understood. This chapter presents college-level classroom examples illustrating how students rush to apply techniques when given statistical tasks, and examines the implications of students' lack of conceptual understanding for needed assessment practices.

In Chapter 8, Schau and Mattern posit that understanding of conceptual connections (i.e., knowing and understanding the relationships between different statistical concepts and techniques) is necessary for students to be successful in statistical problem solving. They argue that an explicit goal in teaching statistics is to assist students in gaining such understanding, and discuss ways for assessing college students' understanding of the interrelationships among concepts, with a particular focus on the use of concept maps.

Part III: Innovative Models for Classroom Assessment

The chapters in this part illustrate new or improved ways to assess aspects of statistical knowledge in specific contexts of instruction. Several chapters focus on assessment of knowledge elements or skills pertaining to one of the eight common instructional goals listed earlier, while others discuss assessment methods (e.g., using multiple-choice exams or portfolios, assessing projects) which encompass several goals. Regardless of their focus, chapters in this section present diverse and detailed examples for tasks, items, extended problems, or projects that can be used to assess students in a range of instructional environments and levels. Where relevant, chapters also address general theoretical or background issues which guide both the design as well as the interpretation of student responses, and highlight the instructional benefits that can be obtained from using specific techniques or approaches to assessment.

Watson discusses in Chapter 9 issues in assessing students' ability to interpret probabilistic and statistical concepts which appear in the media. The chapter suggests a hierarchy of learning goals for judging outcomes of instruction in this area, provides examples of viable assessment tasks and analysis of student responses based on items from the media, and discusses the implementation of media-based items in ongoing classroom assessment.

In Chapter 10, Curcio and Arzt focus on the problems involved in assessment of statistical knowledge when students work in groups. The authors describe the design and application of an instrument to assess the graph comprehension of middle school students working in a small-group setting. Based on a theoretical problem-solving framework, the assessment instrument is explained and examples for students work are provided to highlight the categories sampled by the instrument. Issues involved in assessing the contribution of group processes to the problem-solving process are also examined.

Starkings offers in Chapter 11 practical advice for teachers who want to use projects in their courses. Project work is a method of allowing students to use what they have learned in statistics lessons in a practical context. This chapter describes the process of designing and completing a project and introduces a method of assessment in stages that gives students an indication of their progress on a project and induces them to continue with the work. Examples of projects are given, two different models for assessment of project work are described, and teachers' experiences with these assessment models are described. The project models and examples described have been extensively used with students of 14 to 18 years of age, but can be adapted for younger or older students as well.

The assessment of project work is also the topic of Chapter 12, by Holmes, though this chapter focuses on assessments done by external examiners, as opposed to assessments conducted by a classroom teacher. In secondary schools in England and Wales, at the end of two years of study of mathematics or statistics, many students undergo a formal assessment of their projects by external examination boards. As a departure from the traditional written examination papers used in the past, this assessment attempts to uncover students' deep understanding of statistics. The assessment methods used to evaluate projects and issues involved in using external examiners are described.

The use of a portfolio assessment in a statistics class is discussed in Chapter 13 by Keeler. The portfolio is a purposeful collection of student work that exhibits the student's efforts, progress and achievements over time. It displays the products of instruction in a way which challenges teachers and students to focus on meaningful outcomes and provides a context for guidance and critique. The use of portfolios is explained and evaluated here in the context of a graduate level statistics course, yet the processes involved in developing portfolios and using them to support learning can easily be generalized to different educational levels. The author addresses how an instructor determines the critical knowledge to be assessed in this format, the learning principles used to guide the assessment process, and the considerations involved in deciding how to implement portfolio assessment and how to rate and score student work.

In Chapter 14, Lajoie describes a way in which computer technology can be used to assess as well as extend statistical learning. The chapter describes a project for eighth grade students in which a computerized library of exemplars (based on audio and video recordings) was created and used to illustrate samples of average and above average performance on statistical projects. A case is built for how technology can be used to provide detailed information, beyond that which is normally provided by traditional paper and pencil tests, on how students progress over time in their statistical problem solving and reasoning.

The first six chapters in this section describe ways that can help capture multiple aspects of students' ability to apply or extend their emerging statistical knowledge and problem-solving skills. However, some of these approaches require that teachers break out of traditional forms of instruction as well. The remaining two chapters take a fresh look at the assessment challenges facing teachers when they have to continue to rely on more traditional forms of paper and pencil tests. Such assessments may be called for to evaluate mastery of computational procedures or of certain statistical techniques, or may be inevitable due to classroom realities related to class size or conditions of instruction.

In Chapter 15, Jolliffe aims to help teachers develop their own assessment instruments for formative purposes through exemplars of written classrooms tasks. Unsatisfactory tasks are used to illustrate the pitfalls, and alternative versions are given as examples of good practice. Some comments on grading are included. The emphasis is on written assessment in the classroom, mainly of pupils aged about 14-19, but much is relevant also to introductory statistics courses at college and university level. Consideration is given to ways of assessing factual knowledge, the ability to use computers, understanding of concepts and application of techniques, and communication skills. The pros and cons of multiple choice and open-ended questions are discussed as are the challenges of oral assessment and assessment of group work.

In the last chapter in this section, (Chapter 16), Wild, Triggs, and Pfannkuch discuss effective uses of multiple-choice items in large introductory statistics college-level classes. These authors acknowledge that many of the methods and ideas presented in this volume about "alternative" assessments, while promising and important, tend to be very demanding in terms of teacher-time and other resources, which are scarce in many academic teaching contexts. It is essential to find ways to employ inexpensive traditional methods more creatively to assess and reinforce attainment of key goals in large classes. Wild et al. discuss the pros and cons of multiple choice items and describe strategies for writing items that address multiple teaching goals, including both the ability to plan and carry out investigations and choose appropriate analyses, as well as interpret what these analyses tell about the world. Conclusions are formed about what can and cannot be accomplished with forced-choice testing.

Part IV: Assessing Understanding of Probability

Over the years, numerous studies have examined understanding of probability, chance, and processes with multiple outcomes using children as well as adults. These studies reveal that many students and adults find such topics difficult to learn and understand in both formal and everyday contexts and that learning and understanding may be influenced by ideas and intuitions developed in early years. Both research and classroom observations often suggest that there may be a gap between the formal computations and probability rules taught and (presumably) learned in many classrooms (e.g., regarding the outcomes of certain chance events, such as flipping coins several times), and the informal understandings and beliefs that students have before instruction and may still possess after instruction.

In this volume, three chapters illustrate selected topics and dilemmas and describe possible approaches to some assessment challenges in different populations and instructional contexts. To be sure, what is considered an assessment challenge is contingent on a teacher's instructional goals, on the extent to which the teacher finds it important to link the teaching of probability with the teaching of data analysis and any other area of statistics or mathematics (such as combinatorics), and on his or her resulting teaching methods. While some teachers may emphasize the learning of the logic behind and rules for probability computations, others may be more interested in making sure students acquire "correct" intuitions about chance processes, or improve their subjective estimates of probability. Overall, the goals for student learning and the assessment of students' understanding of probability are complex and involve many different issues whose full coverage requires a separate volume.

In Chapter 17, Metz examines the nature and assessment of two core ideas that underlie students' understanding of probability in a real-world context: randomness and chance variation. Metz argues that these "big ideas" should serve as instructional goals and that they involve both conceptual construction and beliefs about the place of chance in the occurrence of events in the world. The author considers how young students' ideas or beliefs are acquired within or influenced by the culture of a mathematics classroom, and examines assessment of students' understanding and application of chance in a classroom context, with attention to cognitive constructions, beliefs, and classroom culture.

The role of combinatorics in teaching and learning probability and variables that influence students' work and cause errors when solving combinatorial problems are examined in Chapter 18 by Batanero, Godino, and Navarro-Pelayo. Based on a theoretical perspective about desired knowledge of combinatorics in the context of learning statistics, the authors present different tasks for assessing combinatorial reasoning of secondary school students and examine issues in the interpretation of student responses.

Finally, to shed light on yet another and different context for teaching and assessing probabilistic knowledge and reasoning, Cohen and Chechile examine in Chapter 19 issues in assessing students' understanding of probability in a software-assisted environment. These authors describe methods for assessing students' understanding and interpretations of probability and sampling distributions while using instructional software in a college statistics course. Ideas about identifying concepts embodied in such instructional software and issues in designing questions to test concepts are described. The authors illustrate students' interpretations of computer displays of probability distributions and analyze students' errors to highlight the unique issues involved in assessing knowledge developed through computer applications.

IMPLICATIONS FOR FUTURE RESEARCH AND DEVELOPMENT NEEDS

Many of the chapters in this volume struggle with two overarching challenges: how to create tasks which can tap students' status with regard to one or several of the instructional subgoals described earlier, and what kinds of interpretations can validly be applied to students' responses or behavior (e.g., students' knowledge, reasoning processes, communicative skills, and dispositions). While some assessment tasks may be relatively context-free, the majority of chapters assume that assessment (and of course learning) activities should be embedded in contexts that are meaningful to students. Thus, at all levels of instruction teachers need to deal with a third challenge, that of the establishment of meaningful contexts for teaching and assessment.

Creating meaningful contexts is crucial yet not trivial, as problems have to be arranged so that their conditions and parameters are understood and shared by both learners and their teachers. For younger students as well as college-level students, what seems meaningful or realistic to a teacher may not be so to the student. Furthermore, problems and tasks used in assessment should be designed to reveal how students approach, model, and reason statistically about a given situation, as opposed to how they apply routine procedures.

Although the chapters in this book provide a thoughtful discussion of many current assessment issues and offer many practical suggestions for statistics educators and researchers, they only begin to address the full range of existing challenges in assessment in statistics education. As more attention is focused on statistics education and as more is known about the processes involved in acquisition of statistical knowledge, skills, and dispositions by students at different levels, we hope to see additional areas for assessment explored in more depth, such as:

Assessment of students in computer-assisted environments

This involves two separate subtopics: searching for effective ways to assess what students can do and how they reason when they use computers or other technological aids, and the nature of and limitations on inferences that can be drawn from assessments when students learn with computers but are tested without computers, as is common in many classes.

Assessment of "statistical literacy"

This would involve testing the application or transfer of student learning to interpretive or functional tasks such as those encountered in media or outside the classroom. The challenge in assessment of statistical literacy is that it should involve examining not only what students can do and how they think when asked to, for example, reflect on a report in the media, but also their tendency or disposition to do so without being cued.

Assessment of students' understanding of "big ideas"

Throughout their education students encounter important statistical ideas such as variation, error, bias, sampling, or representativeness. These ideas underlie much of students' overall understanding of the uses and limitations of statistics, yet may not have direct mathematical or functional representation. Their meaning may depend on the context in which they are invoked. Tasks are needed that can assess students' understanding of and sensitivity to the prevalence and importance of such "big ideas" in different contexts.

Assessment of students' intuitions and reasoning involving probability concepts and processes

This area has seen much research activity and there are clear indications that many students have misconceptions or intuitive beliefs that are not being changed during instruction. There is a need to transfer and adapt promising assessment methods and instruments used by researchers (mostly involving in-depth clinical interviews with selected students) to formats that are reasonably acceptable and accessible to teachers and that can be used for "routine" classroom use. Some preliminary work has been done to develop paper and pencil instruments to assess statistical reasoning but these are difficult to validate using traditional measurement approaches.

Assessment of outcomes of group work

A common teaching format in statistics, especially at the precollege level, is the use of groups, especially for project work. However, the assessment and grading of the outcomes of students' work when it is done in groups has been described as the most frequent stumbling block for novice statistics teachers who participate in workshops emphasizing active learning and cooperative group activities. Promising approaches for assessing group work are being developed and implemented in mathematics education and should be examined for their relevance to statistics education.

A broader challenge, and one pertaining to all areas of instruction in statistics, has to do with the need to develop models and methods for comparative assessments of student learning. We need reliable, valid, and practical assessment instruments for measuring and evaluating the relative utility of instructional approaches or curricula used in teaching statistics, or for monitoring the overall level of statistical knowledge or statistical attitudes and dispositions in a certain student population. As long as statistics items used in large-scale or standardized assessments remain focused on computations (as opposed to statistical reasoning) and provide little context, the relative effectiveness of statistics courses or units will remain difficult to ascertain. Furthermore, there will be little chance to change practices of teachers who tailor their instruction to the content of standardized tests. The ideas presented in several of the chapters in this volume can be used to advance the area of comparative assessment in statistics but further investment in this area is warranted.

Overall, we believe that the challenges listed above can be adequately addressed only through collaborative efforts involving teachers, statisticians, measurement experts, psychologists, mathematics educators, and technology specialists. The merits of such a collaborative approach are demonstrated by the rich array of perspectives offered in this volume by an interdisciplinary set of contributing authors. It is essential that we continue to jointly pursue improvements in current methods of assessment in order to ascertain that all students can function effectively as citizens and workers in an information-laden, statistically-oriented society.

Part I

Curricular Goals and Assessment Frameworks

Chapter 2

Some Emerging Influences Underpinning Assessment in Statistics

Andy Begg

Purpose

A number of influences impinge on statistics education. This chapter focuses on three of these that are especially noticeable at the K-12 level but also operate to some extent at the college level: a) the changing place of statistics in the curriculum, b) the emphasis on processes in the mathematics curriculum, and c) new ideas about learning and the ways that assessment is viewed in education in general. These, together with the purposes and principles of assessment, are explored because they underpin aspects of assessment in statistics education.

STATISTICS IN THE CURRICULUM

The first influence is the changing emphasis on statistics. It is used in many disciplines and is needed for participation in society, hence it is gaining prominence in education. This is evident from the emergence of statistics departments in universities and curriculum initiatives in schools that incorporate statistics and suggest that its teaching should begin at an early stage (NCTM, 1989; Australian Education Council, 1991). These curricula emphasise *statistics and probability for all,* which contrasts with the past focus on mathematically mature students in high school and college or those who needed statistics for work in other disciplines.

Traditionally curricula have been subject-based. Now curriculum frameworks are being developed (Ministry of Education, 1993) where two other dimensions are considered: the *essential learning areas* and the *essential skills*. Learning areas (language, science, technology, mathematics, social sciences, arts, health) may be thought of as subjects but are intended to provide a framework for all that is taught in schools. If one considers a matrix with learning areas juxtaposed against school subjects, one can see how the subject mathematics contributes to learning areas (science, technology, social sciences, ...) and how subjects (chemistry, geography, economics, industrial arts, home economics, ...) contribute to the learning area mathematics. With this perspective, when we talk of statistics within mathematics we need to know whether we are talking about the subject or the learning area. The former gives a single-subject focus, while the latter implies cross-curriculum consideration which may be acceptable to teachers of young children but may present a challenge to high school teachers.

Similarly, the essential skills (related to communication, numeracy, information-handling, problem-solving, self-management and cooperation, work and study) can be considered against learning areas or subjects. These skills are generic and can be learned through all subjects, including statistics. They suggest an approach to assessment that emphasises interpretation and communication, the use of computers to access and process data, problem-solving, understanding concepts rather than computation of results, and working with others rather than working individually.

MATHEMATICAL PROCESSES

The second emerging influence is evident from recent curriculum documents (NCTM, 1989; Australian Education Council, 1991; Ministry of Education, 1992). It is the increasing emphasis on mathematical processes (what mathematicians do) alongside content (what mathematicians know), and these processes link closely with the essential skills from curriculum frameworks. In statistics these processes relate to what statisticians do. One listing of the processes is problem solving, reasoning, communicating, making connections, and using tools. These processes are intended to be integrated with content and not treated as separate topics.

Assessment should similarly be concerned with this meld of content and processes. The assessment implied by these processes involves using extended and open-ended tasks that allow students to investigate meaningful problems, reason about data, communicate verbally and write reports, make connections and models, and use computers. It is likely to involve formal and informal observation of oral and written work so that evidence can be collected regarding the quality of the statistical thinking, the understanding of the use of tools, and the connections between situations and their models. To indicate the influence of these processes on assessment, some aspects of each are listed.

Problem solving

Problem solving is more than solving word problems; it includes investigating, modelling and simulating, and doing open-ended project work. It may lead to extended practical projects and assessment needs to reflect this. Some aspects of this are considered by Lesh, Amit, and Schorr in Chapter 6, by Holmes in Chapter 12, and by Starkings in Chapter 11.

In assessing problem solving one would look not only for solutions but for evidence of planning (identifying and exploring problems; formulating plans; collecting, retrieving and analysing data), using a range of strategies (making tables, using graphs, finding relationships), and modelling (using concrete materials, graphs, networks, tree diagrams and flow diagrams; then fitting a model to the data and checking its fit).

Reasoning

Mathematical reasoning is concerned with proof and certainty within given assumptions, but when mathematics moved to problems with multiple solutions this degree of certainty weakened. Now that statistical and probabilistic thinking has been introduced, there are further aspects of uncertainty in reasoning that require consideration.

Aspects one might look for in assessing reasoning include classifying and describing (sorting and organizing data to support an argument); arguing (justifying answers and procedures, drawing conclusions from interpretations, using summary statistics appropriately); inferring (making and evaluating conjectures and hypotheses, interpolating and extrapolating, making appropriate and responsible decisions); and proving (acknowledging assumptions, constructing proofs).

When reasoning is valued the focus for assessment moves from the final answer to the way it was obtained. Ideas about models and the statistical reasoning linked to them are developed by Lesh, Amit, and Schorr in Chapter 6; and by Batanero, Godino and Navarro-Pelayo in Chapter 18.

Communicating

Statisticians are consulted by clients because of their statistical expertise, but their knowledge within a client's field is often limited. Statistical consulting involves finding what the client wants and becoming aware of what extra data may be useful. After collecting and processing all the relevant data, the task is to interpret the results and convey the findings to the client in non-technical terms. Thus, communication is fundamental in statistics and should be reflected in its teaching and assessment.

Communication is also important because it influences the success of teaching and learning programs. It occurs at the inter-personal level (working cooperatively, debating possible courses of action) and at the intra-personal level (understanding what needs to be done, reflecting on and clarifying thinking). It involves listening and speaking (discussing difficulties, asking questions, presenting and explaining results, discussing the implications and accuracy of conclusions, discussing possible interpretations of data, presenting reports); reading and writing (using reference material, writing a report); and representing (using graphs, diagrams, and symbols to represent ideas). While some of these aspects are generic and will be assessed in other subjects, they may also need to be considered within statistics.

Some aspects of communication are considered in more detail, including those related to statistical literacy (Watson, Chapter 9) and projects (Holmes, Chapter 12; and Starkings, Chapter 11).

Making connections

Making connections is important for successful learning and because statistics pervades life and is used in many subjects. Aspects that need to be connected occur within the subject (linking concepts, procedures, and topics within statistics and probability; relating various representations of concepts and procedures; recognising equivalent representations; seeing the subject as integrated), with other curriculum areas (linking statistics with other subjects and with generic skills), and in everyday life (using statistics in work, leisure activities and in familiar and unfamiliar situations; understanding statistical representations in the media). These connections need to be made explicitly, and this linking needs emphasis through encouragement and by being recognised in assessment.

Making connections with real-life problems is considered by Lesh, Amit, and Schorr in Chapter 6; connecting procedures and concepts is discussed in Chapter 7 by Sloane, Kelly and Whittaker; and making links between statistical concepts is a focus in Chapter 8 by Schau and Mattern.

Using tools

Tools are used to gather and process data, to help make sense of them, and to simulate situations. Tools allow computation to be de-emphasised, but if they are used in class then they need to be available during assessment. Elements that one might look for in the use of tools include general uses such as measurement in the collection of data, calculator use for computation, and computer use with applications (word processing, databases, spreadsheets, packages for statistics and graphing, and simulations).

The use of computers in assessment in statistics is discussed further by Lajoie in Chapter 14 and by Cohen and Chechile in Chapter 19.

CONSTRUCTIVISM

The third emerging influence on assessment is from new theories of learning. This issue has been addressed in numerous studies and conferences on assessment in mathematics. For example, in an ICMI study Romberg (1993) gave a general overview of learning, and Galbraith (1993) reported on some of the problems fitting assessment with constructivism. Leder (1992) discussed some similar issues related to constructivism.

Until recently assessment was based on behaviourism. The assumption was that knowledge was broken into small and specific objectives that were transmitted to students, and each objective could be assessed. Behaviourist assessment provided feedback so teachers could make decisions about programs and report on their students. This assessment was usually after learning (summative), and prior knowledge was not seen as important. Some teachers did informal diagnostic (before learning) and formative (during learning) assessment, but because their programs were well defined this did not usually influence their teaching. The focus for behaviourist assessment was on correct responses to questions and on behavioural objectives which usually related to the repetition of taught procedures or the recall of facts. Little consideration was given to understanding concepts, to applying ideas to other subjects and new situations, or to modelling, investigating, or communicating.

The recent move from behaviourism to constructivism requires a consideration of different assumptions. Von Glasersfeld, whose ideas have been accepted by many mathematics educators, assumes that the basic principle about constructivist learning is *knowledge is not passively received but actively built up by learners;* and that *coming to know is an adaptive process in which learners organise their experiences of the world, rather than discover an objective reality* (1989). Teachers influenced by constructivism need to consider what feedback they and their students want from assessment. This feedback may answer a number of questions which arise before, during, and after teaching (Begg, 1991).

Before teaching the questions include:

- What are the student's interests, ideas, conceptions and misconceptions about the related content and processes prior to teaching?
- What are their questions about the topic likely to be?
- What activities could focus on their questions?

These questions suggest informal assessment strategies such as classroom discussion, stimulating students to ask questions, and observing and listening to students at work. Some of the answers to these questions will be known by experienced teachers and the information may be partially gained from teacher-sharing sessions.

During learning the questions include:

- What are the students wanting to know about the topic?
- What processes are they tending to use?
- Are the learning activities focusing on these processes?
- Are the meanings being constructed similar to the intended ones?
- How are the students putting together their ideas?
- Are they developing learning-to-learn skills?
- How might the unit of work be modified to better suit the class?

These questions require the teacher to find out what students are thinking while they are learning. This is likely to be done informally during teacher-student interactions when the teacher is supporting, encouraging, and challenging with "what if" and "what if not" questions that respect the learners' autonomy and their views, but help them move from their incorrect ideas.

Developing learning-to-learn skills suggests that students need strategies to identify what they are learning from the experiences they encounter and to build their ability and confidence with self-assessment. Empowering students to audit their own learning involves helping them to recognise criteria that point to development in learning, teaching strategies for reflection, inviting them to compare their self-assessment with the teacher's assessment, and assisting them in comparing their present knowledge (knowing that, knowing about, and knowing how) with that of others so they can appreciate that other views are possible.

After a topic is covered the assessment questions are:

- What are the students' ideas now and what processes have they been using?
- Are their present ideas and skills different from their earlier ones?
- Can their new ideas and process skills be used in unfamiliar situations?
- What needs to be reported or documented?
- What changes need to be made to the program?

"After learning," or summative assessment, is difficult because it is multi-faceted. It becomes even more difficult with a negotiated curriculum, because, even though activities are chosen by the teacher with specific objectives in mind, these activities are only starting points for open-ended approaches and negotiation with the students. These negotiations determine the direction of the lesson and therefore what learning occurs. Testing for standard concepts will not show all that has been learned in this situation; even in traditional classes it is likely to lead to teaching for tests rather than teaching for understanding. In spite of these difficulties, the teacher needs to know what has been learned and may find this by observation, by marking work, by formal assessment events, or by means such as discussion and concept maps.

PURPOSES FOR AND TYPES OF ASSESSMENT

While the implications of constructivism suggest numerous purposes for assessment, these can be grouped in an overview as two basic purposes: a) for better learning by providing feedback to teacher and learner, and b) for reporting to parents and in-school documentation. It is reasonable to assume that if assessment does not contribute to at least one of these, then it would be better not to assess. Even when it does contribute, if it is time consuming, then one must ensure that its usefulness outweighs the value gained by using the time on learning.

Assessment for better learning occurs mainly before and during teaching. It involves finding what students bring to their learning, whether they are ready for new learning, and what their interests and ideas are. Sometimes assessment for better learning is to provide motivation by building students' self-esteem and by changing their perception of their teacher's expectations of them; in this case if marks are allocated, then they need to be clustered around a high mean.

Assessment for reporting and documentation occurs after teaching. School documentation is important to facilitate the movement of students who change schools and classes and who leave school, and it is needed so that schools can report to parents. These forms of assessment and methods of reporting are influenced by community expectations.

While these two purposes of assessment are generally accepted, the types of assessment that should be used, especially after learning, are more problematic. Behaviourist assessment has usually been norm-referenced with a student's performance compared with that of peers, or criterion-referenced with performance measured against well-defined specific objectives. Wiliam (1994) is typical of educators who see limits with these types, and he suggests that we need to consider two further types—ipsative and construct-referenced assessment. Ipsative assessment is concerned with "value added," that is, the difference between what was known before learning and what is known after. Construct-referenced assessment relates to holistic and open-ended activities such as projects, where it may not be appropriate to define specific objectives or standards. With this type of assessment different students may investigate different topics, but a general set of standards exist from which the assessor selects the appropriate ones for each student's work.

While ipsative and construct-referenced assessment suggest a movement from behavioural assessment, these four types are interrelated. Underlying any criteria are normative assumptions made about what learners of a particular age might generally be expected to know. Underlying "value-added" assessment is the need to find out what was known before and after, and this is likely to involve criteria. Criterion-referenced and construct-referenced assessment relate to external standards, the difference being whether these relate to explicit objectives or whether they involve a choice of objectives for each particular student's work.

PRINCIPLES OF ASSESSMENT

Regardless of the types of assessment used, an agreement by educators should guide the use of the principles of assessment. These principles relate to consistency, focus, range of techniques, individuality, self-assessment, reporting, openness, and practicality.

Consistency

Assessment should be consistent with its purposes (which need to be explicit), and with one's view of statistics and learning. Tasks should yield valid results about learning, and what is assessed should reflect what is important in statistics. What is important is not only facts and skills but includes the processes previously discussed. Thus assessment should involve a range of tasks and require the application of a number of ideas rather than relying on tests which focus on narrow sets of skills such as the correct application of standard algorithms. Assessment should promote equity and avoid inequities such as those related to gender, ethnicity, language and culture.

Focus

Assessment should focus on what students know and can do rather than on what they do not know or cannot do (Cockcroft, 1982), and on the process of coming to know. This suggests tasks should be accessible at numerous levels because all students will not be working at the same level. The tasks also should be more than computational so that alternative conceptions and students' thinking are identified.

Accepting a range of abilities implies having a range of achievement levels. These could be hierarchical—definitions, basic facts, and standard procedures at the lower level; making connections and problem solving next; and statistical thinking and reasoning, communication, and generalisation at the highest level. A preferred approach uses three categories of goals—content, process, and thinking goals, because all students will achieve in some way in each area.

Range of techniques

The limitations of written tasks with students who are not good readers or writers but have verbal facility are obvious, but less obvious are problems with multiple-choice questions where some learners may be more inclined to guess correct answers while other learners may reflect more carefully on options and, with good reasons, think of other factors that make the correct solution problematic. To overcome such limitations a range of techniques is needed. This may include informal tasks (questioning, observing, listening, reading students' writing), written tasks (multiple-choice, matching lists, short answer, completing statements, open-ended questions, short essays, open-book tests, two-stage tasks), oral tasks (presentations, seminars, interviews, debates, role plays, interviews), practical tasks (investigations, projects, simulations, experiments, use of computers and calculators), and cooperative tasks (group activities, peer assessment). A selection of such tasks could form the basis for an assessment portfolio.

Individuality

Assessment should allow for the unusual because the unique aspects in students' responses often help show what sense they have made of the experiences provided and what strategies they have used. It is not unusual for a student to discern a pattern, devise a strategy, or come up with an idea that has merit, which the teacher has not encountered. For a teacher there is fascination and reward in detecting the unusual. An example of this is where a student assumes that a probability is conditional while the teacher assumes it is not, and the student's thinking needs to be probed before the response can be assessed.

Self-assessment

Self-management skills and the aim of establishing an ethos of life-long learning have implications for self-assessment. Teachers can facilitate this by providing lists of topics to be covered, understandings to be gained, an indication of time for various tasks, the criteria being used for assessment, and opportunities to self-assess with marking guides.

Students may use learning logs to record learning, effort, and enjoyment. Such logs may be confidential or be read by the teacher but this needs to be agreed upon at the start. Young children and older students can describe what they have learned from an experience and what they would like to explore. This is particularly useful if there is an emphasis on learners establishing their prior ideas about a topic, identifying their strengths and weaknesses, considering the ideas and strategies of others, and monitoring their own changes. Learning logs may be used to stimulate reflection. Within logs self-assessment sheets might be used and be headed, "Things I have found out," "Things I now know," and "Actions I took to sort out my problems." A section for students' remaining questions may be included and be headed, "Things that still puzzle me" or "Things I would still like to know." For younger children "talking heads" might be used:

I think I understand it　　I am coming to grips with it　　I need help!

Reporting

Reporting provides students and parents with an indication of progress, and Cockcroft (1982) makes it clear that "a cross is of little assistance to a pupil unless accompanied by an indication of where the mistake has occurred and an explanation of what is wrong or a request to consult the teacher when the work is returned." Similarly a pass/fail, a grade, a percentage mark, or a ranked position does not give the same amount of information as a sentence or two written about the student.

The student's right to privacy needs to be considered; some countries have a "privacy" act to ensure this. Feedback for students and parents should not be public. This means that teacher feedback should be by written comment or by a quiet word rather than said openly in class.

Openness

The assessment process should be open to review and scrutiny by students and parents. Students should know what is expected, what criteria are used, when deadlines occur, and how results will be used. If a student misses an assessment deadline then the teacher should give a "best estimate" of the student's work, and not a zero as this is unlikely to stand scrutiny in terms of what the student knows.

Practicality

The assessment program must be practical. It should not be too time consuming for the students or for the teachers, yet it needs to provide information for the teacher to improve the learning program and feedback for students and parents.

IMPLICATIONS

The focus of this chapter was not to discuss assessment alternatives as such, but to make explicit a basis on which assessment can be built. From the issues considered a pattern emerges which supports the approaches introduced in later chapters. These issues fit with ideas from the National Council of Teachers of Mathematics (1995) in the Assessment Standards and are advocated by writers such as Stenmark (1989, 1991), who introduced many teachers to authentic assessment, performance assessment including investigations and writing, observations and interviews, and the use of portfolios. The approaches fit with statistics education because the holistic nature of the subject is taken into account, content and processes are assessed, feedback needed to implement programs based on constructivism is produced, and the approaches fit with the purposes and principles of assessment. These alternatives are discussed in the chapters that follow, in particular, authentic assessment is discussed in Chapter 3 by Vos and Colvin, written tasks by Joliffe in Chapter 15, and portfolio assessment by Keeler in Chapter 13.

Bringing about change is not easy, and if teachers are to change, they require time, support, and encouragement. The first step is for teachers to see assessment in statistics education as problematic and to identify it as an area of their concern. They may start by thinking about any of the issues discussed; alternatively, they may start with a dissatisfaction with their present practice and be looking for specific techniques, but hopefully at some stage these teachers will distance themselves from the specifics and consider the ideas that underpin them.

Chapter 3

Authentic Assessment Models for Statistics Education

Shirley Colvin and Kenneth E. Vos

Purpose

Authentic assessment is an emerging field within assessment models. It claims to measure by direct means the student performance on tasks that are relevant to the student outside of the school setting. Most educators will agree with the need to assess learning within the context of applications. This chapter will address the following issues:

- A vision of an effective assessment system must be articulated. What are the standards (visions) for an effective assessment system?
- A well-thought-out plan for designing an effective program must be constructed. What are the components of a process for designing an effective authentic assessment program?
- Classroom teacher readiness to change assessment plans is crucial to any program of assessment. How do you determine the degree of readiness of classroom teachers for a new assessment plan?
- Promises abound in the assessment field but limitations can strangle an assessment program at conception. What are the promises and limitations of recent assessment reforms?

A crucial aspect of teaching and learning is knowing what and how much is learned. Assessment should be the source of this information. This chapter will give a glimpse of how to design an authentic assessment plan in statistics education.

INTRODUCTION

At the forefront of reform of education and its many inherent issues is how to assess the product of schooling, which certainly includes statistics education. The recent literature points to the need for assessment measures that (a) require the student to transfer learning to new situations/view situations from different perspectives for problem-solving purposes (National Council of Teachers of Mathematics, 1989), and (b) have instructional utility (NCTM, 1995; Webb, 1993; Stenmark, 1991). Therefore, the field of education is actively searching for "alternative" assessment programs. These alternatives can be of many forms but most commonly they are given the label of performance, authentic or real-world application. This chapter will focus on authentic assessment models but obviously it is not possible to present an authentic assessment model without incorporating performance or real-world applications. The three terms associated with assessment, performance, real-world applications and authentic, are sometimes used interchangeably to describe the same type of assessment tasks. However, in this chapter the term, authentic assessment, is used to describe the means of measuring student performance on tasks that are relevant to the student outside of the school setting. The term, performance assessment, relates to the direct, systematic observation of student actions related to a performance task. The term, real-world, is sometimes overused in education to describe any aspect of life outside of school. In order to be consistent and reflect the belief for the need to communicate clearly, this chapter will use the term authentic assessment. At the present time, a concise definition of authentic assessment is being formulated. An acceptable description could be the direct examination of student performance on significant tasks that are relevant to life outside of school (e.g., Herman, Aschbacker, &Winters; 1992). The use of real-world situations in assessment settings is promoted by the NCTM Assessment Standards (NCTM, 1993) and other reform documents (NCTM, 1989; Mathematical Sciences Education Board, 1993). In some ways, the most important aspect of assessment is not the struggle with the correct assessment term to describe the phenomenon but rather the struggle to construct an effective, efficient and meaningful assessment system in an educational setting. The term, system, is used to describe both an individual teacher designing an authentic assessment program for classroom instruction and a more elaborate program of assessment designed by school districts or states. Therefore, a system can be fairly complex but it does not necessarily need to be if a teacher wants to begin a small assessment program in a single classroom setting.

Authentic assessment is a viable vehicle to assess a student's understanding of statistics education. However, authentic assessment is a fairly recent phenomenon in the assessment field. Many issues need to be resolved or clarified before authentic assessment will impact traditional testing situations in American schools. Classroom teachers are excited about the potential of moving from traditional multiple-choice tests to an authentic assessment model. However, the excitement wanes when the reality of this enormous undertaking emerges. Issues of appropriate item development, technical measures of reliability and validity, effective scoring designs, and acceptance by the public to new reporting schema quickly appear. These issues should be viewed as challenges within the field of assessment and not barriers. A clear and concise linear model of dealing with these issues may not be possible but a process of thinking about how to effectively design an assessment system can be developed. This chapter examines issues involved in creating and using authentic assessment tasks as part of a broader system of assessment in statistics education. The first section, Task Issues, focuses on the nature of appropriate assessment items and on scoring of students' performance. The second section, System Issues, focuses on principles for designing an effective authentic assessment system and reiterates known limitations and barriers to success. The last section, Implications, discusses the resulting challenges and the potential for changing and improving how assessment occurs in various statistics education settings.

TASK ISSUES

Authentic assessment tasks must involve activities appropriate to a student's life outside of school. One way to judge if a task is authentic is to apply the Reality Principle (see Chapter 6). If the student believes the situation within the task could really happen in "real life," it is more likely the student will attempt a possible solution. A significant challenge for anyone considering the possibility of writing an authentic assessment task is the need to capture the interest of the student, make the situation believable to the student, incorporate sufficient real data to lend credibility and, most importantly, focus the process toward important statistical concepts. If the authentic assessment task does not engage the student, a major part of the goal of the assessment task is not satisfied. If the student doubts the situation occurs in "real life" outside of school, a reluctance to continue to explore in depth may occur. If the situation incorporates real data but it seems to be too simple in form or accuracy, a student will quickly question its authenticity. If the assessment task is engaging, believable, incorporates sufficient real data but obviously lacks the need to use any important concept in statistical education, then the assessment task should be redesigned or eliminated from the assessment program.

The following examples can be considered authentic assessment tasks. Each example lists an appropriate age range but with adaptation, considering the development level of the student, the example could be extended to older or younger students. A few examples of authentic assessment tasks will be given in a brief outline format which includes a general comment, grade level(s), important ideas, and some techniques needed. A more detailed example adapted from *Measuring Up* (1993) completes the list of examples and serves as a context for elaborating on scoring issues in authentic assessment.

Who Fits My Rule?

"Who Fits My Rule" is classification guessing game in which players try to figure out the common characteristics, or attributes, of a set of animals or other living objects.
Grade levels: primary--kindergarten through third grade
Important ideas: collecting, organizing and interpreting categorical data
Some techniques needed: classification schema (e. g., Venn diagrams, visual displays, charts)

Fit-all Mitten Company

A small group of students (3 or 4) constitute a project group responsible for an action plan answering at least the following questions posed by the management council of the Fit-all Mitten Company.
    What size mittens will reflect the company's name "Fit-all?"
    What color or colors should the Fit-all mitten be if limited yarn colors are available?
    What style (closed or glove) Fit-all mitten should be produced?
Grade levels: middle, junior high, early high school
Important ideas: collecting information from a limited sample, organizing the data in visual displays, making decisions based on acquired information
Some techniques needed: designing a method of obtaining information, displaying information, determining hand size by an approximation method

Heart Disease

While the cause of heart disease is not known there are many risk factors associated with heart disease. Have either individual or groups of students choose a factor or combination of factors to investigate. Use information currently available from a variety of sources.

Grade levels: upper high school, introductory college
Important ideas: significant issue to society, pose complex questions using data from external sources (surveys, experiments)
Some techniques needed: multiple representations, causation and relationship factors

Heavy Bears

The core of the example comes from *Measuring Up* (1993) pp. 125-132.
Grade levels: advanced 3rd, 4th grade and above
Important ideas: use real-world context for data, organizing unordered data in a meaningful way, compare data sets
Some techniques needed: drawing a line plot, analyzing sets of data, choosing appropriate representations

The data in Table 1 give the weights of some grizzly bears and black bears living in the Rocky Mountains in Montana. Some of the employees of the Montana agency responsible for natural resources in the state believed a study needed to be done to verify the cubs (young bears) of different types of bears maintained the relationship of weight and type of bear found in older more mature bears.

1. Organize these data in a way that would help you find which kind of bear is heavier—grizzly or black bear?
(More space allotted on student response sheet)
2. Write down three things that you can tell about the weights of the bears.
(More space allotted on student response sheet)
3. Based on these data, how much heavier is a typical bear of one kind than a typical bear of the other kind?
Show how you figured out your answer.
(More space allotted on student response sheet)

Table 1. Weights of Bears

GRIZZLY			BLACK		
Bob	male	220 lbs	Blackberry	female	230 lbs
Rocky	male	170 lbs	Greta	female	150 lbs
Sue	female	210 lbs	Freddie	male	140 lbs
Linda	female	330 lbs	Harry	male	230 lbs
Wilma	female	190 lbs	Ken	male	170 lbs
Ed	male	180 lbs	Hilda	female	220 lbs
Glenda	female	290 lbs	Grumpy	male	160 lbs
Bill	male	230 lbs	Blackfoot	female	150 lbs
			Marcy	female	170 lbs
			Grempod	male	200 lbs

Quality standards

The data set must be viewed as a whole in order to allow comparison of the two groups of bears. Since these data are unordered there must be evidence of displaying these data to show the overall shape and features of the data set. The data sets must be described using three different statements. A summary of these data must show the comparison of the two data sets, not just comparing the heaviest and lightest bears.

Scoring rubric

A general scoring rubric as well as a specific scoring rubric are shown in Figure 1. The general scoring rubric is applicable to many similar situations while the specific scoring rubric applies only to this assessment task. Therefore, the general scoring rubric is a model for future specific scoring rubrics of other assessment tasks.

General Scoring Rubric	Specific Scoring Rubric
3 points • used reasonable strategy(ies) to reach a conclusion • gave clear explanation • complete explanation or only a minor error	3 points [HIGH] • clear, accurate graph or plot • description includes range and distribution comments • grizzly bears are heavier than black bears • value reported is difference between central values of grizzly/black bear data sets
2 points • used reasonable strategy(ies) but did not finish or reach a conclusion • gave unclear explanation • some deficiencies • incomplete	2 points [MEDIUM] • inaccurate display of data • description does not include range and center of data sets • concluded all grizzly bears are heavier than all black bears • incomplete explanation of typical bear comparisons
1 point • started a description but unsuitable statements • major errors • inappropriate approach	1 point [LOW] • graph or plot has major flaws • description focuses on individual bears rather than features of the whole data set • no typical data compared
0 points • blank or unreadable • incorrect with no logical explanation	0 points • no explanation • random statements about bears

Figure 1. Scoring Rubrics

In *Measuring Up* (1993, pp. 130-32, permission granted for use) the results of three different 4th grade students are reported. These results are reproduced in Figure 1. The results were scored using the Specific Scoring Rubric with High being 3 points, Medium is equivalent to 2 points and Low is 1 point. The example of the High score included an accurate line plot with both medians shown. The display is designed to assist in supporting the conclusion that young grizzly bears weigh more than young black bears. This response met or exceeded most of the High (3 points) criteria. The example of the Medium score shows the young grizzly bears weigh more than the young black bears but the horizontal axis is not meaningful in this situation. It assumes the bears are numbered or ordered from least heavy to most heavy in weight. This display does not include a description of the centers of the data sets. The example of the Low score shows a display of all the weights of the bears without distinguishing between grizzly and black bears. It is a display of the information but has major flaws since it focuses on individual bears rather than on the comparison between the types of bears.

After reviewing a few tasks it is appropriate to step back and consider an entire assessment program within a classroom, department, or school district. The word chosen to describe this type of assessment program is an assessment system. The next section gives a brief introduction to standards for an effective assessment system, appropriate development phases for assessment task construction and some impediments to assessment reform.

SYSTEM ISSUES

Standards for an effective assessment system

An assessment system can only be considered effective if it is set in a context of a vision of what constitutes sound practice in assessment. The word, standard, is gaining stature and importance in the current curriculum reform movement. Unfortunately, with its vaulted stature has come multiple meanings. A standard can be a vision or goal or benchmark to reach in the near future or it can be a hurdle to jump before considered successful. In this chapter a standard is a vision or benchmark used to measure the appropriateness of the assessment system. It is not a hurdle to jump over or slide under. These standards follow the general format and substance of other compilations of standards (NCTM, 1995; MSEB, 1993). Each standard can be viewed separately but the impact of each standard is greater when all six standards are considered together. These six standards are general principles for beginning the process of designing an assessment model. This general principles section is followed by development phases which describe the more practical steps taken to design an assessment model for the classroom, department or school district. Both the general principles and the development phases lead into a section on impediments to assessment reform.

General principles

1. Important Statistics Education Content. The assessment should reflect the statistics content that is most important for the students to learn. Important statistics content should be stated as completely as possible. Agreement obviously may be difficult to attain. However, there exists at least a core of big ideas agreed to among statistics educators.

2. Enhanced Learning Of Statistics. Assessment should intensify the learning process by a student. In other words, the assessment should add to the learning of statistics and not be viewed as a necessary evil, e. g. test to confuse. Ideally learning and assessment should be a seamless process.

3. Development Levels Of Learners. The development levels of learners should be reflected in the assessment. Some areas of mathematics education such as geometry and number theory have stated the specific stages of development by learners. Statistics learning does not possess a precise schema. However, general knowledge of the development of learning from the early years through adult should be used to match the appropriate assessment task with the development level of the learner.

4. Criteria For Performance. The assessment plan should include a process to determine how the criteria of performance are developed. Experience working with classroom teachers has shown that a very difficult aspect of an assessment system is how to establish sound criteria of performance. There must be stated a flexible process in determining who sets the criteria, how the criteria are set, who implements the criteria, how are the criteria revised, and how are the criteria reported.

5. Multiple Sources Of Information. Assessment should include multiple situations and models. An assessment system must incorporate more than a single source of information. Authentic assessment tasks alone are not rich enough to give a complete picture of individual students or groups of students. There needs to be a diversity of assessment sources such as multiple-choice test, authentic assessment tasks, and portfolios to name just a few. Recognizing the importance of multiple sources of information is most crucial to the success of an assessment system.

6. Openness. All aspects of the assessment process should be open to review and scrutiny by the public, teachers, learners, and administrators. The use of the results should be clearly understood by everyone involved. Openness in assessment can be very elusive. One can believe in openness for an assessment system but practice just the opposite. This is a common pitfall for many assessment systems. Openness is necessary for support from the various groups who have a stake in the assessment results. Many assessment systems neglect to communicate the process or clearly state the results in a readable manner.

Development phases

On a practical level, the process of designing an authentic assessment system is not linear, concise, or prescriptive. However, there are phases which should be considered in the design. A group of instructors or an individual teacher could follow these phases to begin to understand how authentic assessment could impact the classroom. One can start at the first phase and continue in a linear fashion or begin at different entry points and skip around in the process. The general principles are a net over the development phases. The general principles do not match one-to-one with the phases but rather are just that: principles. The following process has proven to be successful with many classroom teachers over the last few years.

Identify the important ideas in statistics education	*Note*: the number of important ideas should be small (6-10) in order to focus the content and assessment.
Identify knowledge, skills, and techniques needed to understand the important ideas	*Note*: avoid too finely detailed statements, rather keep the framework as broad as possible.
Explore how the important ideas in statistics are used in society/business/education	*Note*: "authentic" means related to life outside of school, not just look like real-world.
Design a limited number of authentic assessment situations at different grade levels for pilot use	*Note*: suggested grade level ranges--K-4, 5-8, 9-12, college which matches NCTM *Standards* ranges.
Create quality standards for each authentic assessment situation	*Note*: a quality standard is a statement that describes a successful solution to the task. Also it can be a description of features to consider in a successful solution.
Using the quality standards develop a specific scoring rubric for each authentic assessment task	Note: experience suggests the scoring rubric be a scale with an odd number (excluding zero) of values, not exceeding six values.
Devise a method of recording and reporting learner results in an accurate manner	*Note*: authentic assessment scoring schema are usually not easy to quantify. A concise plan of documentation must be developed.
Conduct a quality review of the entire process	*Note*: quality review means every decision is eligible for revision, nothing is a given or obvious.
More questions, more ideas	*Note*: similar to the process of data analysis, there should be more questions after going through the process than before beginning the process.

Applying the development phases to yield authentic assessment tasks is both a process and obviously a product. An authentic assessment task should be forthcoming at the completion of the phases. The next section includes a few barriers to success in implementing an assessment system.

Some impediments faced by assessment reform

Designing an authentic assessment program for a classroom, department, or school district should be a straightforward process. However, experience and anecdotal evidence does not support this assumption. Many barriers and impediments are inherent in schooling and seem to slow down any changes in assessment programs. These barriers can be grouped into two broad categories: conditions for readiness by instructors and promises/liabilities. The conditions for readiness give a non-exhaustive checklist to use with teachers to gauge their acceptance of a new assessment plan.

Conditions for readiness

1. Desire for better assessment information. Change is not born out of contentment. If teachers are satisfied with the present assessment activities, they will resist any suggestion of moving toward authentic assessment or any other assessment model.

2. Staff openness to innovations. New assessment models can be threatening to many teachers-- by the way, not only "deadwood" teachers can react in this manner.

3. Conceptual clarity about assessment. There is a certain amount of fuzziness currently associated with assessment. Teachers must be convinced about the conceptual underpinnings of any assessment model available.

4. Assessment literacy of staff. Few teachers consider themselves to be assessment experts but they must have at least a working knowledge of assessment. Fulfilling this condition may be the first priority of any assessment plan.

5. Clarity about learning goals. Assessment is not possible with non-existent or out-of-focus learning goals. Before any classroom, department or school district can develop assessment tasks, it must develop or articulate learning goals for statistics education. Failure to fulfill this condition will sink the entire assessment plan. The common learning goals described in Chapter 1 together with age-specific learning goals discussed elsewhere (e.g., NCTM, 1989) could serve as a starting point for a local discussion of learning goals for students in a particular group or level.

6. Openness of community/parents to new methods of assessment. The obsession with "single number" success ratings will be difficult to modify. Many parents are comfortable with the traditional assessment system and may resist any change which seemingly gives less information.

Promises and potential liabilities

An authentic assessment program holds out promises as well as liabilities. Any assessment model should be reviewed for potential liabilities and promises. A balance of promises and potential liabilities should be the goal of an assessment program that hopes to make an impact in the classroom, department or school district. The following chart summarizes some of the promises and liabilities of assessment reform plans.

Promises	Potential Liabilities
Emphasis on critical thinking	Authentic assessment is not equated with complex, higher order problem solving ability
Written communication of mathematical and statistical information highly regarded	Emphasis on a specific mode of communication assumes a certain learning style
High stakes testing more closely aligned to statistics education curriculum	"Opportunity to learn" statistics becomes paramount
New challenges in technical quality e. g. validity, reliability	Conventional quality measures for testing may not be feasible
Classroom teacher an integral key to the assessment reform	Heavy dependence on teachers for item design, field testing of items, and scoring

IMPLICATIONS

An authentic assessment model is a natural for assessing the learning of statistics. Statistics is making sense of data drawn from authentic settings. To fully understand the power of statistics a student must be able to "make sense" of data. An individual, school or college may use the standards of an effective assessment system, but should always be aware of barriers that may impede the assessment models used.

Authentic assessment holds an alluring promise for many important features of statistics education: critical thinking emphasized; written communication highly regarded; close alignment to statistics curriculum; ability to make decisions based on data. However, to successfully implement an authentic assessment system there must be articulation of the vision of an effective assessment system, building on this vision with a process of designing an assessment system, but always with a keen awareness of the barriers to success in implementing this assessment system.

The success of any assessment system depends on a carefully constructed plan from conception to implementation. It is possible to make authentic assessment be a success in our schools. Success in statistics assessment must be measured by the increase in learning and understanding the power of statistics. Only time will tell if statistics understanding of our students will increase within the authentic assessment era.

With a renewed interest in a balance among curriculum reform effort, new instructional strategies, and alternative assessment techniques, educators are faced with a difficult task of teaching and learning in the 1990s and in the next millennium. Navigating through this maze of contradictions and unfilled promises, educators need to at least keep this focus: It is better to try at least one different assessment technique than be paralyzed with all this information and refuse to try any new assessment program. Experience has shown that trying at least one new assessment program will open a new dialogue of understanding between a student and the classroom teacher. If this goal of opening a dialogue is met with attempting authentic assessment, we would call the effort a tremendous success!

Chapter 4

Monitoring Attitudes and Beliefs in Statistics Education

Iddo Gal, Lynda Ginsburg and Candace Schau

Purpose

Students' attitudes and beliefs can impede (or assist) learning statistics, and may affect the extent to which students will develop useful statistical thinking skills and apply what they have learned outside the classroom. This chapter alerts educators to the importance of assessing student attitudes and beliefs regarding statistics, describes and evaluates different methods developed to assess where students stand in this regard, provides suggestions for using and extending existing assessments, and outlines future research and instructional needs.

INTRODUCTION

"I was terrified when I learned that I would have to take [statistics] because I have always had a mental block dealing with mathematical formulas."

"I have found math to be easy for me throughout school. I think statistics would be fun."

"My teacher said statistics can be misleading and in any case do not relate to people as individuals."

"Although I have never taken a statistics course, I hear they are very difficult and abstract."

These comments, written by high school and university students who had not learned statistics before, show that students may enter statistics education, at either the secondary or post-secondary levels, with strong feelings or ideas involving this subject. A central tenet of this chapter is that, while teachers of statistics are focusing on transmitting knowledge and skills, students may be having an easy or difficult time learning or applying statistics due to the attitudes and beliefs they carry with them. Despite the apparent importance of this topic, little has been written in the professional (i.e., research- or teacher-oriented) literature about possible ways to measure students' feelings, attitudes, beliefs, interests, or expectations in the area of statistics, and about issues and dilemmas involved in such assessment.

This chapter is organized in six parts. Part one presents a rationale for attending to belief and attitude issues by those teaching statistics (especially in secondary and college contexts). Part two surveys some definitional and background issues. Possible sources for students' beliefs and attitudes are examined in part three. Part four presents basic approaches for assessing statistics attitudes and beliefs, both for the purpose of illustrating what instruments are "out there," as well as to help clarify the various facets or components of statistics attitudes and beliefs. Additional ways to extend the range of information obtained in a survey of students' attitudes are introduced in part five. The last part discusses the current state of the field and outlines implications for practice.

WHY CONSIDER ATTITUDES AND BELIEFS IN STATISTICS EDUCATION?

Students' attitudes and beliefs regarding statistics deserve attention for three reasons: 1. their role in influencing the teaching/learning process (process considerations); 2. their role in influencing students' statistical behavior after they leave the classroom (outcome considerations); and 3. their role in influencing whether or not students will choose to enroll in a statistics course later on, beyond their first encounter with statistics (access considerations).

Process considerations

Increasingly, one of the stated goals of statistics education at all levels is to develop flexible statistical problem-solving, statistical literacy and related communication skills, and data-analyzing skills, as opposed to merely imparting computational and procedural skills (NCTM, 1989; Moore, 1990). The creation of a problem-solving environment for learning statistics requires teachers at all levels to build an emotionally and cognitively supportive atmosphere where students:

Feel safe to explore, conjecture, hypothesize and brainstorm and are *not afraid* to experiment with applying different (statistical) tools and methods,

Feel comfortable with temporary confusion or a state of inconclusive results as well as the uncertainty inherent in statistical and probabilistic situations,

Believe in their ability to navigate or "muddle through" intermediate stages, temporary roadblocks, and the decisions needed to reach a certain goal; and

Are *motivated* to struggle with and keep working on tasks or problems which may require extended investment of energy.

However, many students are not ready to embrace and function within a problem-solving-oriented learning environment in statistics education. Part of this lack of readiness is due to the attitudes they carry from their experiences with mathematics (and mathematics teachers). Statistics teachers should be able to assess and monitor students' feelings and ideas, so as to make sure all students either have or develop the dispositions described above and required to function effectively in a problem-solving environment, and to detect those students who develop unproductive beliefs or negative attitudes, so that some assistance can be offered.

As students and later as teachers of statistics courses at the post-secondary level, the authors have seen and felt how attitudes and beliefs, especially negative ones, can have a direct impact on the classroom climate and on individual students' opportunity to learn. Strong negative emotional responses such as crying are obvious manifestations of student distress, but the presence of even one student with continuing negative attitudes in a class can create an uncomfortable atmosphere. Similarly, strong positive responses (for example, "aha–I've got it!") help create a positive climate.

Outcome considerations

Most students take a first (introductory) course in statistics either at the precollege level, or as a "terminal" elective or compulsory course at the college or graduate level, i.e., they will not have to take another course unless they wish to. (In addition, some students enter a quantitatively-oriented program, such as in statistics or business, where an introductory course is followed by more advanced statistics courses later on.)

Within the first two contexts, which include the majority of those learning statistics, two goals for an introductory course in statistics can be posited (see Gal & Ginsburg, 1994 and Chapter 1 in this volume):

- To prepare students to deal effectively with statistical situations in the world outside the classroom, and have the know-how as well as the dispositions needed to act as a smart citizen or consumer in a modern society.

- To prepare students to handle, use, or interpret research or statistical data in their academic or professional discipline.

These goals imply that students should emerge from statistics courses with a willingness and interest to think "statistically" in relevant situations. Teachers should aim to engender in students a positive view of statistics and an appreciation for the potential uses of statistics in future personal and professional areas relevant to *each* student.

Access considerations

Some students may have to take further statistics courses, after their first college course, in order to complete a program of advanced or graduate studies. In this regard, an introductory course should provide the foundations for understanding more advanced statistics. More importantly, students' early encounter with statistics should be positive, so as not to prevent otherwise promising students from entering a program with quantitative requirements due to their negative attitudes towards statistics or negative beliefs in their ability to understand statistical topics.

Overall, process, outcome, and access considerations in statistics education imply that it is incumbent upon statistics educators to know what are students' attitudes and beliefs towards statistics before, during, and after taking a statistics course.

DEFINITIONAL AND BACKGROUND ISSUES

There is a definitional challenge in discussing students' ideas, reactions, and feelings about statistics and learning statistics. Though the terms "attitudes" and "beliefs" have been frequently used by researchers and teachers in this regard, little explicit attention has been paid to the distinction between them. Researchers, for example, have often implicitly defined statistics' attitudes or beliefs as whatever their favorite assessment instrument measures. The extensive body of research on affective issues in mathematics education can be used to guide a discussion of affective responses to statistics education.

In conceptualizing the affective domain of mathematics education, McLeod (1992) distinguishes among emotions, attitudes, and beliefs. Emotions are fleeting positive and negative responses triggered by one's immediate experiences while studying mathematics or statistics. Attitudes are relatively stable, intense *feelings* that develop as repeated positive or negative emotional responses are automatized over time. Beliefs are individually held *ideas* about mathematics, about oneself as a learner of mathematics, and about the social context of learning mathematics that together provide a context for mathematical experiences. This description of attitudes and beliefs seems compatible with other research from within the social science field that explores affective and cognitive components of students' beliefs or attitudes regarding a school subject (Green, 1993; Edwards, 1990; Millar & Millar, 1990).

Applying McLeod's terminology to statistics education, we focus on beliefs and attitudes, rather than on emotions, which are transient and hard to measure but important in that they can be intense and serve as a source for development of attitudes.

Beliefs that would be important to consider by those involved in statistics education may include, but are not limited to:

- Beliefs about mathematics (e.g., is it hard/easy, requires innate skills, it can be mastered by anyone),
- Beliefs about the extent to which statistics is part of mathematics, or requires mathematical skills (e.g., statistics is all computations)
- Beliefs about what should happen or transpire in a statistics classroom, or expectations as to the culture of a statistics classroom (e.g., a lot of drill and practice with textbook problems, a lot of talking about real-world examples)
- Beliefs about oneself as a learner of statistics or mathematics (e.g., I am good at it, I don't have what it takes)
- Beliefs about the usefulness or value of statistics and its importance in one's future life or career (e.g., I will never use it and don't really need to know it)

Together, students' web of beliefs along these interrelated facets provides a context for their approach towards and interpretation of classroom experiences in statistics (and mathematics). Beliefs take time to develop and cultural factors play an important part in their development. They are stable and quite resistant to change, with a larger cognitive component and less emotional intensity than attitudes.

Attitudes towards statistics represent a summation of emotions and feelings experienced over time in the context of learning mathematics or statistics. They are quite stable with moderate intensity, and have a smaller cognitive component than beliefs. Attitudes are expressed along a positive-negative continuum (like-dislike, pleasant-unpleasant), and may represent, for example, feelings towards a textbook, a teacher, a topic, a project or activity, the school, etc.

While the discussion above helps to distinguish between beliefs about and attitudes towards statistics and its learning, these conceptual clusters are clearly related. As McLeod (1992) argues, attitudes influence and are influenced by one's own beliefs. Therefore, below we unite the discussion of beliefs and attitudes, but address them separately where necessary. We use the inclusive phrases "attitudes and beliefs regarding statistics" with the understanding that there may be several targets to attitudes or facets of beliefs, as described above, and that students' beliefs and attitudes may differ for each target or facet in this regard.

SOURCES OF STUDENTS' ATTITUDES

Many of the existing research studies identify and measure attitudes toward or beliefs about statistics in students who have just enrolled in an introductory college level statistics course. What could be the sources for students' attitudes, if this is their first encounter with the discipline? Three sources for attitudes and beliefs are examined below.

First, students may have had previous experience with statistics in school-related contexts. This experience could have occurred through reading or doing research that uses statistics. Some students may have completed some statistics education before college; we expect the number to increase as the NCTM Standards (1989) are increasingly implemented in K-12. In addition, many students drop out of college level introductory statistics courses (Del Vecchio, 1994, found a drop rate in undergraduate introductory statistics courses at a major Southwestern U.S. research institution of about one-third); when they re-enroll in the course they bring with them attitudes developed during the previously unsuccessful experience with statistics.

Second, most people have "notions" of what statistics means based on their out-of-school lives. One project examined this issue using high school seniors who had not studied statistics in school (Gal & Ginsburg, 1994). A group of twelfth graders from a prestigious private school in the Philadelphia area, all of whom were college bound and in the process of applying to high-level universities or colleges, were asked, "What do people study when they take a statistics course? What comes to your mind when you hear the term statistics?" The following quotes illustrate the range of responses obtained:

"When I hear the term statistics, I usually think of basketball statistics (% of shooting, number of rebounds, etc.) or survey statistics (as in 40% of teenagers hate peanut butter)."

"I'm not exactly sure what people study when they take a statistics course, but I think it's along the lines of percentage and graphing, etc."

"Numbers and figures of surveys come into my head. I think of people having a boring life if they make a profession of it, because I know it's a lot more complicated than what I said."

"Although I have never taken a statistics course, I hear they are very difficult. Being a huge sports fan, when I think of statistics I think of how many goals or touchdowns someone has."

"I imagine a statistics course as boring and factual as math. Statistics are gathered data (1000 people live in PA), information good for newspapers, writers, and lawyers."

"Statistics is when someone takes the scores of many things, such as baseball statistics. Math is used a great deal in finding statistics."

Some of these statements contain elements that reasonably portray part of what actually happens in statistics classes and when people use statistics, but in others the information is tenuous or incorrect. Since almost all of the items on most attitude surveys include the word "statistics," it is important to realize that some high school or would-be college students convey some fuzziness regarding what the term "statistics" might be about or about life domains where statistics may be used. How this "fuzziness" affects the validity or usefulness of surveys of precollege students is thus a matter for some concern.

The third possible source is that students believe statistics is mathematics and so their attitudes toward mathematics are merely transferred to statistics. Several of the high school students' quotes, as well as the college students' quotes presented at the beginning of this chapter, support the strong presence of mathematics issues in statistics attitudes.

Schau (see Schau, Dauphinee, & Del Vecchio, 1992) asked students in graduate-level statistics classes to briefly describe their feelings regarding mathematics and statistics and regarding courses in these disciplines, including why they thought they had these feelings. Two general themes emerged: teacher (and class) characteristics, and achievement. At the beginning of the classes, students almost unanimously attributed positive feelings to satisfying past achievement, usually in mathematics. For example, students wrote:

"At elementary school I excelled in arithmetic and this gave me the confidence to tackle areas of mathematics that were more challenging."

"For some reason math was easy to me as I was growing up."

However, students attributed negative attitudes at the beginning of classes to poor teaching coupled with poor mathematics self-concept and achievement. For example, students wrote:

"I had a couple of [mathematics] teachers that were sarcastic and I would feel stupid and helpless."

"In algebra, I found the teacher impossible to understand and eventually gave up."

"I had the same instructor for Algebra I and II and geometry. His methods of teaching included public humiliation if one did not understand the material."

The above informal findings suggest that those planning assessment of students' attitudes in statistics should discriminate aspects of attitudes towards statistics from attitudes towards mathematics. Also, assessments should seek to understand not only the range of attitudes and beliefs, but also their sources, as the same reported attitude may have different bases.

BASIC METHODS FOR ASSESSING ATTITUDES

A literature review of published literature and conference proceedings generated descriptions of nine instruments whose authors claim measure attitudes toward statistics; all use statements for which respondents mark their agreement or disagreement on 5-point or 7-point Likert-type scales. The most commonly mentioned and used include:

Statistics Attitude Survey (Roberts & Saxe, 1982),
Statistical Anxiety Rating Scale (Cruise, Cash, & Bolton, 1985),
Attitudes Toward Statistics (Wise, 1985),
Survey of Attitudes Toward Statistics (Schau, Stevens, Dauphinee, & Del Vecchio, 1995; Dauphinee, Schau & Stevens, 1997).

These four key instruments yield in between one and six scores on subscales that are supposed to represent distinct aspects of statistics attitudes. Dauphinee (1993) provided a thorough description and evaluation of many of these surveys. Two of them, the ATS and the SATS, are briefly described below.

A widely used survey is Steven Wise's (1985) Attitudes Toward Statistics scale (ATS). Wise developed the ATS to correct limitations he found in Roberts' Statistics Attitude Scale (see also Roberts and Reese, 1987), the only attitude survey existing at that time. The ATS uses 29 items to measure attitudes in two areas. The *field* scale (20 items) aims to measure a student's beliefs about the value of learning statistics and the use of statistics in his or her chosen field of study. The *course* scale (9 items) aims to measure affect associated with learning statistics and attitudes toward the course in which a student is enrolled. Example items (one positively and one negatively worded) include:

Field
Statistics is a worthwhile part of my professional training.
Studying statistics is a waste of time.

Course
I would like to continue my statistical training in an advanced course. (this item is the only positively worded one in this scale)
The thought of being enrolled in a statistics course makes me nervous.

After reviewing the existing statistics attitude surveys and considering post-secondary research and instructional assessment needs, Schau et al. (1995) determined that a good survey should exhibit seven important characteristics. It needs to include scales (and items) that (1) measure key aspects of statistics attitudes, (2) are based at least partly on input from introductory statistics students, since they will complete the survey. The survey should be (3) applicable in most post-secondary introductory statistics courses, (4) work both at the beginning of and throughout a course with only tense changes, and (5) be short and so minimally disruptive when administered in class. Scales should include items that (6) measure both positive and negative attitudes. Finally, (7) the number of scales and the items that constitute them should be supported when confirmatory statistical techniques, such as confirmatory factor analysis, are applied to students' responses.

Accordingly, Schau et al. (1995) developed the Survey of Attitudes Toward Statistics (SATS), which consists of 28 seven-point Likert-type items measuring four aspects of post-secondary students' statistics attitudes. The SATS has two forms, a "pre" form for students who have not yet taken a college statistics course, and a "post" form, for administration during or after a course. The items on both forms are identical except for some wording changes related to the timing of assessment. The availability of two forms enables comparison of attitude and belief patterns at different points in time in the learning process. Appendix A includes all items of the "post" version and scoring instructions.

Below are descriptions of the SATS scales, with examples of items, one positively and the other negatively worded, from the "pre" form. (A comparison to the "post" items listed in Appendix A will show the wording differences required due to timing of the assessment.)

Affect (6 items measuring positive and negative feelings concerning statistics):

   I will like statistics. (item 1)
   I am scared by statistics. (item 21)

Cognitive Competence (6 items measuring attitudes about intellectual knowledge and skills when applied to statistics):

   I can learn statistics. (item 23)
   I will have no idea of what's going on in statistics. (item 9)

Value (9 items measuring attitudes about the usefulness, relevance, and worth of statistics in personal and professional life:

   I use statistics in my everyday life. (item 12)
   I will have no application for statistics in my profession. (item 19)

Difficulty (7 items measuring attitudes about the difficulty of statistics as a subject):

   Statistics formulas are easy to understand. (item 4)
   Statistics is highly technical. (item 26)

Scores on these scales vary in their interrelationships. Scores on *affect* and *cognitive competence* are strongly related to each other. Scores on the *value* and *difficulty* scales are moderately related to those on the *affect* and *cognitive competence* scales but are not related to each other. The internal consistency of each of the scales is at least adequate, ranging from above .6 to above .8. Because statistics students and instructors identified these four aspects of statistics attitudes, and because a confirmatory analysis likewise identified 4 factors, Schau and her colleagues (Schau et al., 1995; Dauphinee et at., in press) believe that the SATS measures four important aspects of attitudes toward statistics.

In addition to the instruments described above, Green (1993) developed a paper-and-pencil modified semantic differential assessment of students' attitudes toward statistics. Some researchers (e.g., Pretorius & Norman, 1992) simply revised existing mathematics attitude surveys by changing the word "mathematics" to "statistics" throughout all items, yielding an instrument that does not necessarily capture unique aspects of statistics.

Other measures of attitudes and beliefs related to statistics were developed as part of research and development projects, mainly those funded by the National Science Foundation. One example is a 25-item, Likert-type instrument developed in 1990 as part of the Statistical Reasoning in the Classroom Project at the University of Pennsylvania (Gal, 1993) for use with high school students and with adults at large. It was later modified for use in evaluating the ChancePlus Curriculum (Konold, 1990). Garfield (1996) used a modified 10-item version in course evaluations for the CHANCE project. This abbreviated form, dubbed SCAS (STARC-CHANCE Abbreviated Scale) and attached in Appendix B, illustrates an omnibus approach to assessment in a classroom context, where testing time is very limited. The form uses single items to enable some coverage of each of several issues related to outcomes of statistics instruction. The use of single items instead of longer scales reduces the reliability of assessment. Yet, it enables detection of students with extreme scores and identification of broad changes in students' response patterns, thus being of some value for both teaching and evaluation purposes.

EXTENDING ASSESSMENTS OF ATTITUDES AND BELIEFS

It was earlier argued that statistics educators would need instruments for at least two key tasks: to perform initial assessment of students' attitudes, and to monitor changes in these attitudes during and after statistical education experiences. In addition, attitudinal measures may be used as part of summative assessment of the effectiveness of a statistical education experience. In all cases, the assessment results should inform preventive or remedial interventions or at least provide increased instructor and student awareness of attitudes.

It appears possible to use existing instruments such as the ATS or SATS in a limited fashion to meet these assessment goals at the college level. When used with secondary or younger students, response possibilities can be collapsed to three possibilities (such as NO, DON'T KNOW, and YES) or represented with a continuum of faces showing sad expressions for Disagree, neutral faces for Neither Agree nor Disagree, and happy faces for Agree (see Begg, this volume). A problem may exist, though, if students are unfamiliar with or have undeveloped notions of the meaning of the term statistics.

Instructors can examine class averages, distributions, or score profiles to determine the status of students' attitudes. If a class scores around or above neutral (e.g., a mean score of 4 on a 7-point scale) on each scale, for instance, the instructor knows that the class as a group does not have an attitude problem. If, however, the class falls much below neutral, the instructor may need to devote more class time to dealing with the negative attitudes. Determining what "much below neutral" means is an instructor-based decision. The scores can be used in a similar way to identify individual students who may have negative attitudes that may interfere with their learning.

However, instruments such as the ATS or SATS have not been designed to provide information about causal factors and about the sources of the attitudes and beliefs expressed by individual students. This information is essential to determine types of support or educational experiences that might be useful for ameliorating students' negative attitudes. Several approaches can be used to obtain the "why" information. Educators can conduct interviews, lead focus group discussions, utilize think-aloud protocols as individuals or groups of students solve problems, or ask students to write journals or histories of their present or past mathematical experiences. Such techniques have been proposed and implemented in the mathematics education field (see, e.g., Tobias, 1993). However, while potentially very informative, they may be impractical in large-class situations or where resources are limited.

To elicit more information about causal factors, instruments such as the ATS or SATS can be extended by adding *open-ended* questions that enable students to describe the intensity and frequency of specific emotional responses (most items normally indicate the existence of a certain attitude, but not how often or how strongly), explain what past experiences underlie their responses, and reflect or elaborate on the causes of their mathematics attitudes, their statistics attitudes, or both. Several methods for extending the basic Likert-type item format are illustrated below.

One approach is to administer a standard attitude survey such as the SATS, ATS, or SCAS, and then ask students to answer in an open-ended fashion the following questions regarding selected items:

Why did you respond as you did?
What aspect(s), if any, of statistics (or mathematics) make you feel this way?
What experiences form the basis for your response?

A related approach is to mark "attitude words" or "belief phrases" in certain items (the examples below use SATS items, one negatively- and one positively-worded), and ask students questions such as the above, with focus on the marked (underlined) parts:

Affect: I am scared by statistics; I will like statistics.

Cognitive Competence: I can learn statistics; I will have no idea of what's going on in statistics.

Value: I use statistics in my everyday life. Statistics is irrelevant in my life.

Difficulty: Statistics is highly technical; Statistics formulas are easy to understand.

Further, it is possible to create open-ended or guided-choice sentence-completion items, possibly based on Likert-type items from existing instruments, and ask students to complete them as well as explain their answers. Items might include (see Gal & Ginsburg, 1994, for more examples):

I think statistics is ... (e.g., useful, interesting, boring, frightening) ...because ...
I think statistics is about... (what topics? What skills?)
I expect that for me, personally, statistics may be later useful for... (write "not at all" if you so feel)
When I think about this course, I'm concerned that ...(write "not at all" if you so feel)

Instructors could also create Likert-type or open-ended items addressing statistical issues based on beliefs identified by Schoenfeld (1992) as typical regarding the nature of mathematics and mathematical activities. For example:

Mathematics (statistics) problems have one and only one right answer.

There is only one correct way to solve any mathematics (statistics) problem - usually the rule the teacher has demonstrated to the class.

Ordinary students cannot expect to understand mathematics (statistics); they should just memorize and apply what they have learned mechanically and without understanding.

Students who have understood the mathematics (statistics) they have studied will be able to solve (address) any assigned problem in five minutes or less.

Other than the commonly-held belief that statistics is heavily mathematical and that statistics is a somewhat difficult discipline, students' beliefs about statistics as a domain remain mostly unexplored. It would be interesting to determine if student beliefs about statistics parallel their beliefs about mathematics. It would also be instructive to examine the profile of scores across attitude scales for each student. Green (1993), for example, suggested that differential profile patterns may affect related behaviors and may be correlated with ease of attitude change.

Finally, completely open-ended questions could be used as part of an initial assessment, in addition to any of the item types presented earlier, to elicit students' responses about broader topics. For example:

Describe any concerns you may have about your educational experience involving statistics.

Describe the extent to which your prior academic background may assist or impede your learning of statistics.

What factors may cause your performance to be poor or good?

How do you feel about learning mathematics or math-related topics in general?

Responses to open-ended questions could be examined informally, or content-analyzed to yield qualitatively distinct response categories and computed percentages of frequent response types. This could be done by the instructor, or turned over to students to form the basis for classroom projects involving data analysis.

IMPLICATIONS

The process, outcome, and access considerations discussed at the beginning of this chapter suggest that statistics teachers should use assessments of attitudes to understand students' presuppositions and, with continuous monitoring, to identify areas of frustration for individual learners, guiding the provision of supportive interventions.

Only a small number of studies (less than 50) and a few instruments (about 10) have been published on the nature and correlates of statistics attitudes, and many of these studies are small-scale or limited. In comparison, a recent review by Helgeson (1993) noted that more than 700 studies have been published on students' attitudes towards science and that more than 50 different instruments have been developed over the years. Reviews of research on students' attitudes, even in established fields such as science education (Helgeson, 1993) or mathematics education (McLeod, 1992) repeatedly point to two problems in the research in those fields: lack of theory-based work, including both construct issues with attitudes and lack of models to guide research, and narrow use of restricted research methods, including a heavy reliance on survey-type measures.

Existing instruments for measuring statistics attitudes are still in an experimental stage and may suffer from the same limitations as those noted in other fields. Yet, we believe that educators cannot ignore the possibility that attitudes, achievement, and persistence influence each other in statistics education in ways similar to those found in mathematics and in other areas. Existing research on attitudes in statistics education points to a small to moderate relationship between attitudes (measured by the ATS scale) and achievement in statistics at the post-secondary level (see Green, 1994; Waters, Martelli, Zakrajsek, & Popovich, 1988; Wise, 1985; Woehlke, 1991). Schau et al. (1992) reported similar relationships between course grade and pre-and post-course attitude scores on *affect*, *cognitive competence*, and *value* scales of the SATS. Del Vecchio (1994) showed a relationship between the *cognitive competence* scale of the SATS and persistence: students who reported more confidence in their abilities to do statistics were more likely to complete their course with a passing grade.

We expect attitude and belief issues in statistics to become increasingly important as more students at all educational levels experience statistical education. For K-12, the NCTM *Standards* (1989) recommend the inclusion of projects and novel problems that take extended time commitments and that may have multiple correct answers. As McLeod (1992) and others (e.g., Meece, Wigfield, and Eccles, 1990) have suggested, these kinds of learning situations are more likely to cause affective responses than the traditional restricted problems types that many students have learned to expect.

With the above in mind, three challenges face the statistics education community regarding the domains of statistics attitudes and beliefs. First, the existing measures of attitudes in statistics are only partially suited to assess all the variables of interest. This chapter has presented various methods to extend existing instruments, based mostly on elicitation of information through open-ended items. The development of assessment instruments capable of providing information relevant for instructional purposes at both the college and precollege levels involves many challenges, both conceptual and methodological. Continued efforts to improve and systematize alternative item formats and examine their reliability and validity are needed to improve the quality and conceptual coverage of the measures currently available.

Second, there is almost no research on the nature of statistics attitudes and beliefs, on their relationship with achievement and persistence, and on attitude patterns in different types of learners (e.g., group differences among males and females or minority and non-minority students). Most designs used by researchers studying attitudes towards statistics show little sophistication, and statistical analyses and presentation of results are often ill-suited to the research questions asked. Research based on longitudinal and other appropriate designs, and employing relevant measures, is needed if we want to better understand how attitudes and beliefs about statistics develop and change throughout encounters with statistics instruction. Such research is needed if we want to develop causal models and be able to plan and test the efficacy of corrective interventions.

Lastly, in order to make the learning of statistics less frustrating, less fearful, and more effective, especially among college students but also at earlier stages, further attention by statistics educators should be focused on the attitudes and beliefs students bring into statistics education experiences, how they develop and change during their educational experiences, and the impact they have on students' achievement, persistence, and eventual application of their new knowledge and skills. This chapter has presented several approaches, both formal and informal, that educators at all levels can use to monitor where their students stand on several facets of beliefs and attitudes towards statistics, its learning, and its use. The use of such methods is paramount if we want to achieve the vision of statistical literacy for all.

Appendix A

This appendix includes the "post" version of the SATS questionnaire, intended for assessing students' attitudes and beliefs towards statistics during or after a statistics class. A "pre" version also exists, for administration before the onset of instruction. The "pre" version includes the same items as the "post" version, with minor wording changes where necessary. It is not reproduced here due to space limits, but sample items are listed later to illustrate the wording changes. Interested readers can contact Candace Schau at the address below, for more information or for the full forms of both versions of the SATS.

Prof. Candace Schau
Psychological Foundations Program
Simpson Hall, College of Education
University of New Mexico
Albuquerque, NM 87131
USA
E-mail: cschau@unm.edu

Survey of Attitudes Toward Statistics (SATS) Post version

DIRECTIONS: The questions below are designed to identify your attitudes about statistics. The item scale has 7 possible responses, ranging from 1 (strongly disagree) through 4 (neither disagree nor agree) to 7 (strongly agree). Please read each question. From the 7 point scale, carefully mark the one response that most clearly represents your agreement with that statement. Use the entire 7 point scale to indicate your degree of agreement or disagreement with our items. Try not to think too deeply about each response. Record your answer and move quickly to the next item.

(Note to the reader: Each of the 28 items below should be followed by a 7-point response scale as described above).

1. I like statistics.
2. I feel insecure when I have to do statistics problems.
3. I have trouble understanding statistics because of how I think.
4. Statistics formulas are easy to understand.
5. Statistics is worthless.
6. Statistics is a complicated subject.
7. Statistics should be a required part of my professional training.
8. Statistical skills will make me more employable.
9. I have no idea of what's going on in statistics.
10. Statistics is not useful to the typical professional.
11. I get frustrated going over statistics tests in class.
12. Statistical thinking is not applicable in my life outside my job.
13. I use statistics in my everyday life.
14. I am under stress during statistics class.
15. I enjoy taking statistics courses.
16. Statistics conclusions are rarely presented in everyday life.
17. Statistics is a subject quickly learned by most people.
18. Learning statistics requires a great deal of discipline.

19. I will have no application for statistics in my profession.
20. I make a lot of math errors in statistics.
21. I am scared by statistics.
22. Statistics involves massive computations.
23. I can learn statistics.
24. I understand statistics equations.
25. Statistics is irrelevant in my life.
26. Statistics is highly technical.
27. I find it difficult to understand statistics concepts.
28. Most people have to learn a new way of thinking to do statistics.

Scoring: Subscale scores are formed by summing the items listed below for each subscale. The scoring for the starred (*) items should be reversed (1 becomes 7, 2 becomes 6, etc.). Higher total scale scores will then correspond to more positive attitudes.

Affect:
1   2*   11*   14*   15   21*

Cognitive Competence:
3*   9*   20*   23   24   27*

Value:
5*   7   8   10*   12*   13   16*   19*   25*

Difficulty:
4   6*   17   18*   22*   26*   28*

Extensions: Depending on the reason for using the SATS, i.e., for instructional or research purposes, additional information can be collected. It may include the respondent's age, ethnicity, degree type, degree status, number of years of high school mathematics taken, number of college mathematics and/or statistics courses completed, computer background, etc. Examples for other attitude related items (note label changes in the scales) that can be used for research purposes in addition to the 28 SATS items include:

How confident are you that you can master introductory statistics material? (A 7-point scale from "Not at all confident" to "Very confident")

How well did you do in your high school mathematics courses? (A 7-point scale from "Very poorly" to "Very well")

How good at mathematics are you? (A 7-point scale from "Very Poor" to "Very Good")

How much experience with statistics (e.g., courses, research studies) did you have before taking this course? (A 7-point scale from "None" to "Great deal")

In the field in which you hope to be employed when you finish school [college], how much will you use statistics? (A 7-point scale from "Not at all" to "Great deal")

Appendix B

SCAS Instrument

The ten items below have been used as part of evaluations of college-level statistics courses. These items do not comprise a scale, i.e., each item is looked at separately. This instrument demonstrates how attitude and belief items can become part of a simple tool for assessing attitudes and beliefs as well for collecting a student's self-appraisal of his or her understanding of statistical issues. Items 5 and 6 directly assess attitudes and beliefs of the kind addressed by the SATS (see Appendix A). However, responses to several other items may also be influenced by students' beliefs.

Scale:	1 Strongly Disagree	2 Disagree	3 Neither Agree, nor Disagree	4 Agree	5 Strongly Agree

1. I often use statistical information in forming my opinions or making decisions.

2. To be an intelligent consumer, it is necessary to know something about statistics.

3. Because it is easy to lie with statistics, I don't trust them at all.

4. Understanding probability and statistics is becoming increasingly important in our society, and may become as essential as being able to add and subtract.

5. Given the chance, I would like to learn more about probability and statistics.

6. You must be good at mathematics to understand basic statistical concepts.

7. When buying a new car, asking a few friends about problems they have had with their cars is preferable to consulting an owner satisfaction survey in a consumer magazine.

8. Statements about probability (such as what the odds are of winning a lottery) seem very clear to me.

9. I can understand almost all of the statistical terms that I encounter in newspapers or on television.

10. I could easily explain how an opinion poll works.

Part II

Assessing Conceptual Understanding of Statistical Ideas

Chapter 5

A Framework for Assessing Knowledge and Learning in Statistics (K-8)

Susan N. Friel, George W. Bright, Dargan Frierson and Gary D. Kader

Purpose

This chapter provides an overview framework for thinking about what teachers and students should know and be able to do with respect to learning statistics at the K-8 levels. Given the number of concepts to be considered and our limited knowledge about the complexities of learning these concepts, we focus on the understanding of graphical representations, examine examples of "good tasks" that may be used to assess graph knowledge, and reflect on what we have learned about the complexities of assessing students' graph knowledge when using these tasks.

INTRODUCTION

What do teachers and students need to know and be able to do with respect to statistics in the elementary and middle grades? Our answer centers on the development of "data sense", which includes being comfortable with posing questions, collecting and analyzing data, and interpreting the results in ways that respond to the original question asked. It also includes comfort and competence in reading, listening to, and evaluating reports based in statistics, such as those found in newspapers, magazines, television, and other forms of popular press. That is, data sense encompasses not only understanding the graphs and statistics that are presented but also evaluating the statistical investigation process used to generate that information from which the graphs and statistics are constructed.

One critical part of data sense is understanding that a statistical investigation is really a process. A statistical investigation typically involves four components (1) posing the question, (2) collecting data, (3) analyzing data, and (4) interpreting the results, in some order (Graham, 1987). Kader and Perry (1994) suggest a fifth stage of a statistical investigation: communication of results. The resulting model gives structure to our understanding of the type of reasoning used in statistical problem solving.

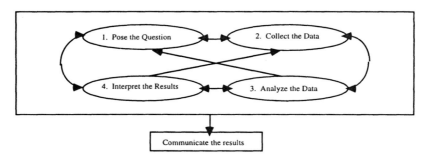

Figure 1. The Process of Statistical Investigation

As with any *process*, there is inherent difficulty in attempting to capture the *quality* of its use by students. Having students do projects that involve the use of the statistical investigation process may be one of the better ways to assess capability, yet such assessment is fraught with difficulties. How do we determine what students know? How do we know that they know it? Part of the answer lies in a better understanding of the components of the process and the related statistical concepts.

This has led us to consider what it means to understand and use graphical representations as a key part of what it means to know and be able to do statistics. The current literature tells us very little about how such knowledge develops. There is some anecdotal and written evidence obtained through developmental research (Gravemeijer, 1994) associated with three different curriculum development projects which helps to frame a set of issues related to understanding graphical representations. Too, graphicacy (Wainer, 1980) in the curriculum, particularly at the elementary and early childhood levels has become a focus of curriculum analysis (Lappan, et al., 1996; Russell & Corwin, 1989).

We have begun developmental research to examine upper elementary and middle grades students' learning of key concepts related to the use and interpretation of graphical representations. We have looked at how such understanding changes over time and with instructional intervention provided by knowledgeable teachers in order to develop a structure for understanding the mathematics that is involved. In the remainder of this chapter, we focus on assessment strategies for the "analysis of data" and "interpretation of results" components of the statistical investigation process that relate specifically to the understanding of different ways to represent data. The focus on data representation in general and graph knowledge more specifically is due to the fact that this is such an important part of being able to use statistics in the real world. Very little seems to be known about students' understanding in this area. Hopefully, learning more about students' understanding of data representation may also allow us to address more general assessment issues for other aspects of statistics.

ISSUES RELATED TO GRAPH KNOWLEDGE

Understanding the process of data reduction and the structure of graphs are factors that influence graph knowledge. The transition from tabular and graphical representations which display raw data to those which present grouped data or other aggregate summary representations is called data reduction. The purpose of data reduction is to identify appropriate representations of the data which remove as much detail from the data as is possible while providing sufficient information to address the specific question at hand.

Graphical representations of numerical data reflect different levels of data reduction. A representation may display the original raw data or grouped data. For example, line plots and stem plots present the original data, while box plots and histograms present grouped data. Most graphical representations used in the early grades (e.g., picture graphs, bar graphs) involve either the original data or tallied data from which the original observations may be obtained. Students in upper grades more often use graphical representations of grouped data (histograms, box plots) from which it is not possible to return to the data in its original form.

In addition to misunderstandings that may emerge with data reduction, the structure of graphical representations of data may also impact understanding. For example, graphical representations utilize one axis or two axes or, in some cases, not have an axis. For graphical representations that use both axes, the axes may have different meanings. In some simple graphs, the vertical axis may display the value for each observation while the vertical axis for more typical bar graphs and histograms provides the frequency of occurrence of each observation (or group of observations) displayed on the horizontal axis. Confusion may develop if the different functions of the x- and y-axes across these graphs are not explicitly recognized.

Another factor that may influence graph knowledge is the ways students are asked to read graphs to gather information. Curcio (1987 and Chapter 10 in this book) conducted a study of graph comprehension assessing fourth and seventh grade students' understanding of four traditional "school" graphs: pictographs, bar graphs, circle or pie graphs, and line graphs. She identified three components to graph comprehension: reading the data ("How many students have 12 letters in their names?"), reading between the data ("How many students have more than 12 letters in their names?"), and reading beyond the data ("If a new student joined our class, how many letters would you predict that student would have in her name?")

SOME TASKS TO ASSESS GRAPH KNOWLEDGE

Informally, teachers can frequently cite instances of confusion that exist when students work with different graphs. Instructional and assessment strategies seem to be needed to help students focus on the characteristics of the graphs, on transitions from one data representation to another, and on different levels of interpretation tied to graph comprehension. Appropriate questions for both instruction and assessment may be developed through application of Curcio's scheme of reading the data, reading between the data, and reading beyond the data. Transitions among data representations may be structured using the transformational strategy of developing bar graphs from work with line plots and histograms from work with stem plots.

There are a number of different kinds of questions used in both written tests and interview settings that we have been exploring with both upper elementary and middle grades students and practicing teachers which provide possible directions for assessment that focus on earlier-noted concerns. With each assessment, we have sought to embed tasks within understandable contexts and to design questions that would call attention to the process of statistical investigation. That is, we have chosen contexts that seem to relate to students' everyday lives. Examples of written problems that we have used as both pre- and post-test instruments with students address the use of line plots and bar graphs and stem plots and histograms (Friel & Bright, 1995).

Assessing First Steps in Data Reduction

One problem undertaken by student involves the context of raisins (as shown in Figure 2).

Students brought several different foods to school for snacks. One snack that lots of them like is raisins. They decided they wanted to find out just how many raisins are in half-ounce boxes of raisins. They wondered if there was the same numbers of raisins in every box. The next day for snacks they each brought a small box of raisins. They opened their boxes and counted the number of raisins in each of their boxes. Students are presented with a line plot showing the information the class found:

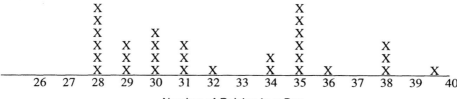

Are there the same number of raisins in each box? How can you tell?

Figure 2. Counts of Raisins

In our research, we grouped students' responses into categories based on the ways they talked about data and the features of the graph, as shown below:

Properties of the graph (considering both range of data and frequency)
- No, because the x's are not all on one number.
- No, because the X's show how many boxes had that many of raisins. Like 28 had 6 and 29 had 3.
- No. If there were the same number in each box there would be X's all above the same number.

Literally "reading" the data from the graph.
- No there aren't the same number of raisins in each box, I found my answer by looking at the data, 6 boxes have 28, 3 have 29, 4 have 30, 3 have 31, 1 has 32, 2 have 34, 6 have 35, 1 has 36, 3 have 38, and 1 has 40.

Properties related to the context or to the data.
- No, because they weigh the boxes until they equal 1/2 ounce. They don't count the raisins.
- No, because some raisins can be smaller and that means you can have more.

Range of the data (considering only range and does not include frequency)
- Because it says the number of raisins goes from 26 to 40.

Frequency of occurrence/height of bars
- No, the X's have different numbers, so there are different numbers of X's in each box.
- No, because some do not have as much X's and some have more.

Other (includes incomplete, unclear, incorrect, or not statistically-reasoned responses)
- No there are not. They all have different amounts.

In this problem, a limited number of students (roughly 28% of sixth graders) were able to reason using information about the data values themselves (from the axis) and the frequencies of occurrence of these data values (the X's). The number of students who seemed to focus on the frequency or number of X's as the data values indicates that there may well be confusion even when using line plots about the role of data values and frequencies. We have found such confusions exist with students' reading of bar graphs; we attribute some of these confusions to having to read the frequency using the vertical axis. For a line plot, this is not the case.

Assessing Next Steps in Data Reduction

Another problem is an investigation undertaken by a middle grades class of students (as shown in Figure 3).

Students were interested in how they used their time. They brainstormed a list of ways such as sleeping, eating, after school sports, and so on. Jim reminded them that some of their time is used just traveling back and forth to school. Some of the students thought this shouldn't count because it really wasn't much time at all. Others disagreed. The class wondered, "What is the typical time it takes to travel to school?"

Students taking the test are told that these students spent time discussing how they would collect their data. For the first look at this problem, they each decided that they would time how long it took them to reach school the next day. They used stop watches (which the teacher had) or timed their trips using their own watches. Once the data were collected, the students made a stem-and-leaf plot (stem plot) for the data number of minutes it took them to travel to school.

Minutes to Travel to School

0	3 3 5 7 8 9
1	0 2 3 5 6 6 8 9
2	0 1 3 3 3 5 5 8 8
3	0 5
4	5

2 | 5 means 25 minutes

1. How many students are in the class? How can you tell?
2. How many students took less than 15 minutes to travel to school? How can you tell?
3. Write down the three shortest travel times students took to get to school.
4. Write down the three longest travel times students took to get to school.
5. What is the typical time it takes for students to travel to school? Explain your answer.
6. Make a histogram to show the information about travel time that is displayed on the stem-and-leaf plot.

Figure 3. Travel times to school

In our research, we found that, after instruction, students were able to correctly answer all four of questions 1-4 which required them to either 'read the data' or 'read between the data' on the given stem plot. Their answers to question 5, which required them to 'read beyond the data', varied somewhat but did show a range of responses correctly utilizing the statistical measures of center they studied. Students were less likely to compute measures of center as part of their responses. Some samples of various types of responses follow:

Responses that identified the mode in the data:
- The typical time it takes for the students to travel to school is 23 minutes, because there are more students in that class that takes them 23 minutes to travel to school.

Responses that identified a cluster of times:
- 10 to 28 because they have most of the data on it. And that is where it is clumped.
- I think between 3 to 20 minutes. Because 15 people are between 3 to 20 and that is more than half.

Responses that provided a tally or range of numbers that occurred most frequently:
- 23-28 min. I know because these numbers are repeated more.
- 3 & 5, you see how many of the same number you can find. [Note: counted digits in leaves]
- Probably 3, 16, 23, 15 & 28 because more people have them 5 numbers.

In the question noted, responses that identified the mode and responses that identified clusters of data are both appropriate; however, how do we evaluate these two responses? Is one "more appropriate" than the other; do we want students to move beyond the use of the mode as a tool in this case to using clusters of data as a way of describing what's typical? If so, what is a "good sized" cluster to be highlighting?

WHAT OTHER KINDS OF TASKS COULD OR SHOULD BE USED IN ASSESSING GRAPH KNOWLEDGE?

Not surprisingly, we can learn a great deal by trying to understand how students explain their thinking about graphs. Although students often give "correct answers" to questions about graphs (as evidenced by the fact that 97% of the students answered the "raisin question" in Figure 2 correctly), the reasoning that supports these answers is often faulty. Students seem to be interpreting graphs in ways that are inconsistent with clear understanding of the underlying mathematics.

Too, a large number of students provide vague or incomplete responses that seem to say, "The answer is this because the graph says so." This level of clarity may reflect the usual emphasis in mathematics on "getting an answer" with little emphasis on follow-up questioning that would create an expectation for clear explanations of why an answer is given. Students may simply not be experienced enough at explaining to be able to it in writing. On the other hand, students' difficulty at explaining may result from our own lack of clarity about how we expect students to be able to talk about graphs and our corresponding lack of clear criteria for what we expect students to say. For example, for question 5 in the "travel time" scenario (Figure 3), it is not clear how we ought to respond to the students who tallied occurrences of numbers. What do such responses tell us about where the students are in their thinking about the concept of "typical" and about what we might want to do to develop this concept over time? In and of itself, we cannot tell if this response reflects an initial approach to answering the question, the result of a full investigation by this student, or something else. It is also interesting that for this question few students used the median or the mean, although these ways for summarizing data had been included in the instruction and would have been an appropriate way to answer this question. What does their inattention to appropriate statistical concepts tell us about their thinking?

Based on even this little information about students' thinking, it is possible to propose other kinds of questions that could be asked to help both us and the students clarify reasoning strategies further or to help focus on both the data values and their frequencies as components involved in reading a graph. The first two examples below deal with how to use data to create a graph. The third deals with detailed interpretation of one small part of a given representation.

1. Someone has opened 5 boxes of raisins and each box has the same number of raisins. How might a line plot that shows these data look? Why?
2. Someone has opened 5 boxes of raisins and two of the boxes have the same number of raisins. Of the remaining 3 boxes, each has a different number of raisins. How might the line plot that shows these data look Why?
3. In the line plot in Figure 2, what do the four X's above 31 mean? Explain.

We do not know a great deal about how to use forced choice items as a way to assess students' graph knowledge. However, we do have models for ways to think about doing this. One model involves students being asked to choose among representations as they read through a number of short problem statements; either students simply choose the appropriate graph (Mathematical Sciences Education Board, 1993) or they choose the appropriate graph and identify certain of the graph elements as well. These examples of forced choice items also require that students justify their reasoning and, in some situations, respond to a short answer interpretive question as well.

Currently, very little research has been done on the usefulness of such items a part of an overall strategy for assessing knowledge. Such problems reflect an attempt to address the problem solving process of reversibility (Rachlin, 1992) in which we move from students working from a problem to data collection to creation of representations; here we have students working from a problem, bypassing data collection and construction of representations, to interpretive reading of an incomplete representation constructed by someone else.

Another model, the use Mystery Graphs (Russell & Corwin, 1989), involves giving students an incompletely labeled graph showing data from a specific context. Students describe what context is being described by the graph. One example is a graph that shows the weights of a number of different lions in zoos across the United States; as part of the data set, there are a few cubs included. This type of problem is not a "fixed choice" problem yet the responses are contained within a realm of prediction about what is possible, given the data as they are displayed in the selected representation.

IMPLICATIONS

Among the reasons for encountering difficulties in assessing knowledge of statistics is the need to clarify what it is that we want students and teachers to know and be able to do with respect to statistics. We currently know very little about the development of many statistical concepts over time and through instructional interventions. Our work in the area of data representations has provided a model of one way to build both understanding based on consideration of the development of graph knowledge and of strategies for assessment that may be used to support and evaluate this development.

Instructional and assessment questions need to provide opportunities for students' thinking to be revealed. Just asking for answers can be misleading for teachers as they try to understand what students know. Sometimes this thinking can be revealed by asking students to explain their answers. But we also need to develop questions (such as the three examples in the previous section) that focus on detail that students might not otherwise talk about or think about. Questions like these have the potential to reveal students' levels of concept attainment as they talk about their thinking. The process of answering these questions may help students expand and refine their thinking. That is, these questions become learning events as well as assessment events.

We need to use a variety of questions so that we can determine whether students are consistent in their thinking. In this way we will be better able to "triangulate" their thinking. At the same time, we are the first to admit that it is often difficult to know what questions to ask to probe students' thinking. It often takes a lot of reflection on our part to even begin to understand what students might be thinking. It is only in hindsight that sometimes we can think what we should have asked to probe a response.

Part of what we need to know about a student's response is whether the basic understandings (e.g., reading the data) are in place. We don't want to leave the impression that there is a tightly sequenced list of prerequisite skills that students need to master in order to answer questions about graphs, but we don't want to confuse lack of understanding about the question with lack of these prerequisite skills.

The use of graphs and other kinds of representations needs to be viewed as part of the process of statistical investigation and not as an end in itself. Some will argue that such study needs to emerge during the process of investigating reasonably "big" problems that engage students in this process; it is not a question of whether we "teach histograms now" but rather that we wait for a need for various representations to emerge out of "big" problems and to be taught within "big" problem contexts. Our work supports consideration of the process, but does address graph knowledge more explicitly as part of the development of the overall process. Part of the reason for choosing such a direction revolves around the need to understand what we don't understand about ways in which the use and reading of graphs may be misunderstood or misinterpreted by students.

In statistics, data reduction is an essential part of analysis of the data; different graphs emphasize different degrees of data reduction. Past instruction and assessment (e.g., often found in school mathematics textbooks) has not demonstrated an awareness of the connections between the process of data reduction and the choice of graphs. Indeed, there appear to be natural transitions between some representations (i.e., line plot and bar graph or stem plot and histogram) that support building the use of multiple representations in a way that may facilitate understanding.

The use of technology as a tool (e.g., computer database, graphing software, or graphing calculators) offers real potential for students in helping to promote exploration of different representations of data as well as the structure of these representations. There has been very little work done in this arena; we don't know a lot about ways that such data display software or calculators support students' understandings. We have seen students use such tools without much thought; they make graphs that are not appropriate for the data or they do not consider exploring the impact of using different graphs or experimenting with scaling. Specifically, we wonder about the fact that when graphing calculators are used, the graphs have no labels on axes or titles displayed with the representations. Also we wonder about the array of implicit "definitions" about the nature of certain kinds of graphs provided by different computer software programs (e.g., *Cricket Graph III, Statistics Workshop, MacStat,* and *Data Insights*). Each program defines the parameters involved in making histograms in very different ways. In addition, each program provides different options with respect to scaling axes and using relative frequencies as an alternative to counts.

There is a need to monitor learners' changes in thinking as they move among ungrouped and grouped data representations. Once we have some knowledge of learners' thinking, are we clear about what attributes of statistical thinking we want to promote and about ways to promote these attributes? How do we rank responses following an instructional unit? For example, what if we want students to begin to understand what the data tells them and also to understand when they are adding their own lens without support from the data? What instructional and evaluation strategies do we use? Do we begin to push students to ask questions about their observations or conclusions such as "Do the data tell you this? Are you making a judgment based on personal experience and reasoning rather than on the data? Do you have the information you need to make a judgment using only the data?"

We need to understand what students understand prior to and following instruction *and* be clearer about how we will judge their responses in light of what we think reflects sound statistical thinking.

Chapter 6

Using "Real-Life" Problems to Prompt Students to Construct Conceptual Models for Statistical Reasoning

Richard Lesh, Miriam Amit and Roberta Y. Schorr

Purpose

The purpose of this chapter is to examine a "model-eliciting activity," based upon a "real-life" problem situation, in which students were provided with an opportunity to construct powerful ideas relating to data analysis and statistics, without explicitly being taught. Student results of this activity will be examined that reveal the somewhat surprising fact that children, even those who traditionally do not perform well in mathematics, can invent more powerful ideas relating to trends, averages, and graphical representations of data than their teachers ever anticipated. The student results shared in this chapter are not unique. In classrooms where we have piloted and refined problems (including the one presented), one common observation is that many of the children who emerge as "most productive" are often those whose mathematical abilities had not been recognized or rewarded by their teachers in the past.

INTRODUCTION: MATHEMATICAL MODELS AND MODEL-ELICITING ACTIVITIES

The use of realistic problems to assess student understanding is a recommended practice in statistics education. We have found that when students develop their own ideas about problems using realistic models, learning is enhanced. In order to make sense of the problem activity, and the accompanying student results, it is important to consider what is meant by a mathematical "model" and the characteristics of "model-eliciting activities." A mathematical model can be considered to be a functioning system for describing, explaining, constructing, modifying, manipulating, and predicting a complex series of experiences. Models are organized around situations and experiences, and people interpret problem-solving situations by "mapping" them into their own internal descriptive or explanatory systems (models). Once the given situation has been mapped into an internal model, transformations within the model can take place which can produce a prediction within the modeled situation. This in turn can lead to further predictions, descriptions, or explanations for use back in the problem situation. Models help us to organize relevant information and consider meaningful patterns that can be used to generate or (re)interpret hypotheses about given situations or events, generate explanations of how information is related, and make decisions based on selected cues and information. These internal models develop in stages. Early conceptualizations or models can be fuzzy, or even distorted versions of later models, and several alternative models may be available to interpret a given problem situation. As can be seen in the student results and interpretations that follow, the children went through several "modeling cycles" in which they reinterpreted the givens, goals, and solution paths. They made modifications and refinements to their models during each cycle so that useful predictions, generalizations, and descriptions could be made for the given problem.

The problem activity described in this chapter was designed according to the following six principles, in order to create the need for students to construct, refine, and extend significant mathematical models. These six principles were developed by expert teachers along with mathematics educators and researchers (for a more complete description, see Lesh, Hoover & Kelly, 1992). While these principles might appear to be rather like "common sense," we have found that many of them tend to be violated by virtually every problem that we have seen in major textbooks and tests. Therefore, in some sense, they are quite radical.

The Reality Principle:

Could this really happen in a "real life" situation? Will students be encouraged to make sense of the situation based on extensions of their own personal knowledge and experiences? Will students' ideas be taken seriously, or will students be forced to conform to the teacher's (or author's) notion of the (only) "correct" way to think about the problem situation?

The Model Construction Principle:

Does the task create the need for a model to be constructed, or modified, or extended, or refined? Does the task involve constructing, explaining, manipulating, predicting, or controlling a structurally significant system? Is attention focused on underlying patterns and regularities rather than on surface-level characteristics?

The Self-Evaluation Principle:

Are the criteria clear for assessing the usefulness of alternative responses? Will students be able to judge for themselves when their responses are good enough? For what purposes are the results needed? By whom? When?

The Model-Documentation Principle:

Will the response require students to explicitly reveal how they are thinking about the situation (givens, goals, possible solution paths)? What kind of system (mathematical objects, relations, operations, patterns, regularities) are they thinking about?

The Model Generalization Principle:

Does the model that is constructed apply to only a particular situation, or can it be applied to a broader range of situations?

The Simple Prototype Principle:

Is the situation as simple as possible, while still creating the need for a significant model? Will the solution provide a useful prototype (or metaphor) for interpreting a variety of other structurally similar situations?

In the section that follows, the problem activity will be presented. This problem was designed to relate to similar employment experiences the students may have had. Next, the corresponding student results, along with our interpretations, will provide evidence of the models and modeling cycles which occurred as a particular group of students solved the problem. In each interpretation we will illustrate the meaning of the particular model and its function in the modeling cycle.

THE PROBLEM ACTIVITY

The activity that follows is part of the PACKETS program for Middle School Mathematics, developed by the Educational Testing Service for the purpose of portfolio assessment, according to the principles described above. This problem was based on a context that was described in a "math-rich" newspaper article that was discussed by the class as a whole on the day before the "Making Money" problem was presented.

Making Money

Last summer Maya started a concession business at Wild Days Amusement Park. Her vendors carry popcorn and drinks around the park, selling wherever they can find customers. Maya needs your help deciding which workers to rehire next summer.

Last year Maya had nine vendors. This summer, she can have only six—three full-time and three half-time. She wants to rehire the vendors who will make the most money for her. But she doesn't know how to compare them because they worked different numbers of hours. Also, when they worked makes a big difference. After all, it is easier to sell more on a crowded Friday night than on a rainy afternoon.

Maya reviewed her records from last year. For each vendor, she totaled the number of hours worked and the money collected—when business in the park was busy (high attendance), steady, and slow (low attendance). (See the table.) Please evaluate how well the different vendors did last year for the business and decide which three she should rehire full-time and which three she should rehire half-time.

Write a letter to Maya giving your results. In your letter describe how you evaluated the vendors. Give details so Maya can check your work, and give a clear explanation so she can decide whether your method is a good one for her to use.

HOURS WORKED LAST SUMMER

	JUNE			JULY			AUGUST		
	Busy	Steady	Slow	Busy	Steady	Slow	Busy	Steady	Slow
MARIA	12.5	15	9	10	14	17.5	12.5	33.5	35
KIM	5.5	22	15.5	53.5	40	15.5	50	14	23.5
TERRY	12	17	14.5	20	25	21.5	19.5	20.5	24.5
JOSE	19.5	30.5	34	20	31	14	22	19.5	36
CHAD	19.5	26	0	36	15.5	27	30	24	4.5
CHERI	13	4.5	12	33.5	37.5	6.5	16	24	16.5
ROBIN	26.5	43.5	27	67	26	3	41.5	58	5.5
TONY	7.5	16	25	16	45.5	51	7.5	42	84
WILLY	0	3	4.5	38	17.5	39	37	22	12

MONEY COLLECTED LAST SUMMER (IN DOLLARS)

	JUNE			JULY			AUGUST		
	Busy	Steady	Slow	Busy	Steady	Slow	Busy	Steady	Slow
MARIA	690	780	452	699	758	835	788	1732	1462
KIM	474	874	406	4612	2032	477	4500	834	712
TERRY	1047	667	284	1389	804	450	1062	806	491
JOSE	1263	1188	765	1584	1668	449	1822	1276	1358
CHAD	1264	1172	0	2477	681	548	1923	1130	89
CHERI	1115	278	574	2972	2399	231	1322	1594	577
ROBIN	2253	1702	610	4470	993	75	2754	2327	87
TONY	550	903	928	1296	2360	2610	615	2184	2518
WILLY	0	125	64	3073	767	768	3005	1253	253

Figures are given for times when park attendance was high (busy), medium (steady), and low (slow).

Student responses and corresponding interpretations

The student responses contained in this section come from a seventh grade "average ability" inner-city classroom. The students worked in three-person teams, with the members being assigned by the teacher. This particular teacher placed an emphasis on portfolio-based assessment; therefore, these students had considerable prior experience working on at least ten projects similar in size to the "Making Money" problem.

For this activity, the students worked at small tables where a "tool kit" was available that included three graphing calculators and other standard classroom tools. The "work station" also included a Macintosh computer with a 12" color monitor, and software for word processing, spreadsheets, drawing, and making geometry constructions. The teacher passed out the problem and told the students that they were to complete their letter describing a procedure for deciding who to hire by the end of the next day's class.

The solution process that follows includes significant segments from a transcript for a group of students whose names were Alan, Barb, and Carla. Most of the graphs that are shown were originally produced using graphing calculators. But, when the teams presented their work in class, they used posters that contained re-drawn versions of their favorite graphs; and, these graphs usually were constructed using a computer-based graphing spreadsheet and a color printer.

{Approximately 5 minutes pass as students read the problem & discuss it.}

Alan: Oh God. We've gotta add up all this stuff. ... You got a calculator?

Barb: They're in here {the toolbox}. ... Here. {she finds two TI-83 calculators in the toolboxes}.

Approximately five more minutes pass while Alan, Barb, and Carla add the numbers in various rows or columns of the table. Since the three students had made no effort to coordinate their efforts, each went off in a slightly different direction. For example, Barb and Carla both added numbers in the first row of the table (which shows the number of hours that Maria worked);

whereas Alan added the numbers in the first column (which shows the number of hours that all students worked during the busy periods in June).

Carla: (looking at Barb) What'd you get? ... I got 159.
Barb: Yep. ... um ... That's what I got.
Alan: I got, let's see, ... 116.
Barb: You punched them in wrong. ... Here, you read them {the numbers} and I'll punch 'em in.
Alan: (pointing to the numbers in the table) 12.5, 5.5, 12, 19.9, 19.5 ...

Interpretation #1: Inconsistent use of a "hodge podge" of several unstated and uncoordinated ways of thinking

This team's first interpretation of the problem was similar to those generated by most other groups. That is, when students first began to work on the problem:

1) They tended to worry most about "What should I do?" rather than "What does this information mean?" Therefore, their first interpretations focused on computation, and the information that was given was treated as though no data interpretation or mathematization was necessary. Also, when computation was done, it nearly always involved only two-item combinations; it did not involve computations of whole rows or whole columns of numbers.

2) They tended to focus on only a small subset of the information, and they tended to focus on isolated pieces of information rather than focusing on underlying patterns and regularities. For example, Alan, Barb, and Carla focused on only the first information that impressed them most. That is, they focused on only the rows or columns in the table that showed the number of hours that each worker worked. This emphasis was not based on a thoughtful selection about which information was most important. It was simply the first information that came to their attention.

3) Their early interpretations seldom consisted of a single coherent way of thinking about givens, goals, and possible solution procedures; instead, they tended to involve a hodge podge of several unarticulated and undifferentiated points of view. That is, different students think in different ways; and even for a given individual, they sometimes switch (without noticing) from one way of looking at the problem to another way. For example, in the transcript that is given here, when Alan finished adding the first column of numbers in the top half of the table, he began to add the first column of numbers in the bottom half; there was no evidence that he noticed that the top half of the table dealt with hours worked and that the bottom half dealt with money earned. In fact, later in the session, Alan tried to subtract data in the top table from data in the bottom table (he tries to subtract hours for dollars, e.g., $690 - 12.5 hours = ?).

4) They tended to focus only on numbers, and ignored quantity types. For example, the quantity "12.5 hours" usually was read as "twelve point five." This emphasizes "how much" but ignores "of what."

Next, Alan, Barb, and Carla spent approximately five minutes calculating the total amount of time that other workers worked. Carla recorded results in the last column of her table. The table of sums that they produced corresponds to the graph shown below.

Interpretation #2: Focusing on total number of hours for each worker

The graph and table shown here focused on only the total number of hours that each worker worked. In presentations of their results, the notions of "seniority" or "willingness to work" were common justifications that students used for emphasizing "hours worked."

Unlike many other groups that produced the preceding graph as part of their final presentations, Alan, Barb, and Carla did not bother to produce the graph shown below.

MARIA	159
KIM	239.5
TERRY	174.5
JOSE	226.5
CHAD	182.5
CHERI	163.5
ROBIN	298
TONY	294.5
WILLY	173

They only produced the table of sums that would have led to this graph. This seemed to be true for several reasons. First, the table of sums that Alan, Barb, and Carla produced was, in itself, enough to enable them to go on to a new and improved way of thinking about the information that was given. Second, at this point in the session, Alan, Barb, and Carla were only using their calculators to operate on pairs of numbers; they were not operating on whole lists of numbers. Therefore, they were not entering data into their calculators (or their computer) in a form that made it easy for them to produce automatic graphs.

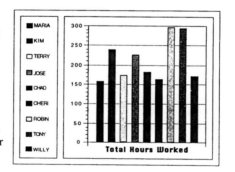

Alan:	OK, so who should we hire? {Alan was looking at Carla's table of sums.}
Barb:	Robin looks good. ... {pause} So does Tony.
Alan:	Maybe Kim. {pause}
Carla:	Hey! We ought to look at money, not hours. ... Money is down here {pointing to the second half of the table which shows the amount of money each student earned}.
Alan:	Yep, money. {Approximately 1 minute passes, as students think and look at the table.}
Barb:	OK, let's add these. (pointing to the rows in the second half of the table).
Barb:	Here, you do Maria. (gesturing to Alan) You do Kim. (gesturing to Carla) And, I'll do Terry.

Alan, Barb, and Carla divided the task into several different tasks, with each working on different parts. In this way, more planning, monitoring, cross-checking, and rethinking tended to occur. Alan, who seemed to be insecure about using a calculator, begins to act as a facilitator and as a monitor for the group, rather than as a person who is actually doing the calculations.

Next, approximately three minutes pass as the students calculate sums in the second half of the table. At this time, Barb becomes the temporary recorder for the group. She takes several minutes to collect the results from the group, and to record these sums in a column (like the one that Carla had constructed earlier).

Interpretation #3: Focusing on the total number of dollars that each worker earned:

Some teams essentially quit working on the problem at this point. For these groups, their presentations often included a graph like the one shown below. ... Again, probably for the same kinds of reasons as for interpretation #2, Alan, Barb, and Carla used only a table of sums; they did not bother to construct the graph shown below.

MARIA	$8,196
KIM	$14,921
TERRY	$7,000
JOSE	$11,373
CHAD	$9,284
CHERI	$11,062
ROBIN	$15,271
TONY	$13,964
WILLY	$9,308

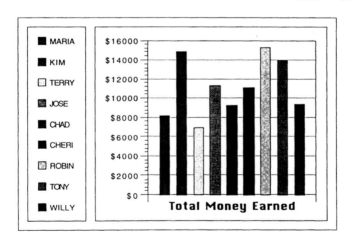

Alan: So, who's the best? ... {pause} ... Robin's best. She got "fifteen two seventy-one." ... And, Kim got "fourteen nine twenty-one." Who's next?
Carla: Tony. ... He got "thirteen nine sixty-four."
Barb: This isn't fair. Some guys got to work a lot more than others. ... Look at Robin and Tony. They worked more than everybody else. That's why they made more money. If Maria worked that much, she'd have made that much money too.
{Mumbling. More than 2 minutes pass.}

At this point in the session, nobody picks up on Barb's suggestion to investigate the relationship between "dollars earned" and "hours worked." Nonetheless, later in the session, Barb comes back to this same suggestion, and at that time, it leads to the idea of investigating "dollars-per-hour" for each worker. Now, however, the students investigate changes in the dollars earned across time.

Barb: Look, Willy didn't work at all in June {pointing to the zeros by Willy's name in the original table.} But, he was doing great in August {pointing to the $3005 by Willy's name in the August column of the original table.} ... Let's just see how much everybody got, totally, in August.

Next, the group spent approximately 10 minutes making a table showing the total number of dollars each worker earned each month. At first, the group only made a list of the totals for August; but, when they were finished with August, they made a table showing all three months. Also, after the values were calculated using calculators, Carla entered the results into the computer spreadsheet. [note: Alan, Barb, and Carla never used the spreadsheet to calculate values; they only used it to record information and to graph results.]

Interpretation #4a: Using a table to focus on the total number of dollars each month.

It is noteworthy that the preceding table was put together in a top-down fashion. Earlier tables were simple lists, and even these lists were created by doing the individual calculations first, and then organizing these results into a well-organized form. The organizational system was not generated first and used as a form to guide the computations that were performed. That is, each of the earlier lists were constructed in a bottom-up fashion.

	Dollars Earned Each Month		
	June	July	August
MARIA	$1922	$2292	$3982
KIM	$1754	$7121	$6046
TERRY	$1998	$2643	$2359
JOSE	$3216	$3701	$4456
CHAD	$2436	$3706	$3142
CHERI	$1967	$5602	$3493
ROBIN	$4565	$5538	$5168
TONY	$2381	$6266	$5317
WILLY	$189	$4608	$4511

Alan: Look at old Willy. He's really catching on {at the end of the summer}. ... Look, back here {in June} he only made a hundred and eighty-nine bucks; but, out here {in August} he was really humming.

Barb: I think August should count most. Then July. ... I don't think June should count much. They were just learning.

Alan: How are we going to do that.

Barb: I don't know. Just look at them {the numbers in the table} I guess. {pause}

Barb: Let's see, out here {in August} Kim was best. ... Then Robin, no Tony. ... Then Robin. ... I think they're the top three. Kim, Robin, and Tony. ... How'd they do in July?

Barb: Wow! Look at Kim. She's still the best. ... But, uh oh, look at Cheri. She was real good in July.
Alan: Let's line 'em up in July. Who's first.
Barb: Kim. ... {pause} Then Tony, and Cheri, and Robin. ... {long pause} ... Then Willy, Chad, and Jose. ... {long pause} ... And, these guys weren't very good {referring to Maria, Terry}.

While Barb was doing most of the talking and overt work, Alan was watching and listening closely. But, Carla was off on her own playing with the computer's spreadsheet, and entering lists of numbers. At this point, Carla re-enters the conversation.

Carla: Look you guys, I can make a graph of this stuff. Look.

For the next four minutes, Carla used the computer to flip back and forth, showing the three graphs that she had made, explaining how she made the graphs, and pointing out who was the top money earner each month.

Interpretation #4b: Using a graph to focus on the total number of dollars earned each month

Similar graphs were made for July and August.

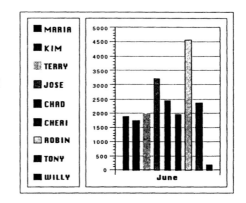

Barb: OK, let's, like, line 'em up for each month.
Alan: You started doing that.
Barb: OK, you {Alan} read 'em off and I'll write 'em down.

For approximately five minutes, Alan, Barb, and Carla worked together to get a list of "top money makers" each month.

Alan: Look, Kim was top in July and August; and, so was Tony. ... Robin was next in August; but, she wasn't as good in July. ... {pause} ... But, she {Robin} was really good in June. ... {pause} ... I think August is most important because some of them were just learning. ... August is how they'll probably do next summer.

Interpretation #5: Focusing on trends in rank across time.

The students noticed that the rankings were somewhat different each month; so, the "trends" shown here were used as an early attempt to reduce this information to a single list.

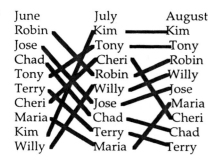

Approximately five more minutes passed while each of the three students nominated workers that they believed should be hired, based on rankings and trends in the preceding table. In most cases, when students spoke in favor of a given worker, they made up some sort of "cover story" to account for the "ups" and "downs" in the performance of the worker. These "cover stories" involved the following kinds of possibilities: (1) some workers learned and improved, while others got bored; (2) some weren't able to work as much as others; (3) some were good during busy periods, but not during slow periods. In these discussions, the students started to pay attention to the fact that the months might not be equally important (e.g., July is the busiest month, August might be the best indicator of current abilities), and that busy, steady, and slow periods might not be equally important (e.g., part-time workers wouldn't be hired during slow periods). In addition, the students began to express concerns about the fact that they would like to have had some additional information that was not available (Who really needed a job badly? Who was willing to work when they were called?). Finally, as Carla is looking at the three-column chart that showed trends (see interpretation #5), she got the idea to make a similar graph using the computer; and this idea leads to interpretation #6.

Carla:	I can make a graph like that {pointing to the table that was used in interpretation #5} with the computer. Want'a see? (see Interpretation #6)
Alan:	Wow! Neat! How'd you do that?
{Carla explains again how she made the graphs using the computer.}	
Alan:	Now who do we pick. ... Who's this?
Carla:	Um, let's see, it's Kim. ... And, this is ... um ... Tony.
Alan:	Who's this?
Barb:	Let me see.
Carla:	Oh, it's Robin.
Barb:	So, we've got Kim ... Tony, and Robin. Who's next? {pause}
Carla:	What about this guy? ... Who is he? ... Um, it's Cheri. ... Look, she was really good here. But, then she screwed up.

Interpretation #6: Focusing on trends in money earned for June, July, and August.

Carla's graph was a line graph showing the total number of dollars that each worker earned for June, July, and August.

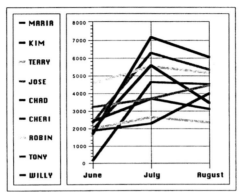

Trends in Money Earned for June, July, & August

Barb: How we gonna decide which of these guys to hire? They were all good some and bad some. ... {long pause} ... How many were we supposed to hire anyway? ... {pause} ... Look at the problem {speaking to Alan}. What does it say?
... {long pause} ...
Alan: We're supposed to hire three full time and three part time.
... {long pause} ...
Alan: I think we should hire Willy. He was good here {pointing to July and August} ... and he didn't get to work much here {pointing to June}.

Interpretation #7: Using telescoping decision rules.

Up until this point in the session, the students implicitly seemed to assume that the best way to choose workers should be to use a single rule for ranking the workers. Then, if this list was successful in ranking workers from "best" to "worst," the top three workers could be hired for full time, and the next three workers could be hired for part time. Unfortunately, no single rule seemed to work to form a single list. For example, both Barb and Carla suggested the idea of using some sort of average. But, this idea was not considered in detail, because the type of averages that were mentioned didn't seem to involve equally important quantities. Therefore, the students began to consider more sophisticated decision-making rules. For example, one rule involved the following kind of two-step process. First-round decisions about who to hire could be based on the ranking in August alone; then, second-round decisions could be based on the ranking in July alone (or based on busy periods alone).

Barb: Look you guys. Some of these people got to work a lot more than others. ... That's not fair. Look, Willy didn't get to work at all back here {in June}.
Carla: So, what're we gonna do?
{Mumbling. More than 1 minute passes.}
Alan: Here. I'm trying something. ... I'm subtracting how much each guy worked. That'll kind of even things out. ... I worked for a guy who did that once. We were cleaning up trash and he wanted us to work fast.

Interpretation #8: Subtracting time scores from money scores

The most important characteristic of this new idea is that, for the first time, it took into account a relationship between the amount of money that was earned and the amount of time that was spent working. But, because the numbers in the tables didn't include any unit labels, nobody noticed that it might not make sense to subtract hours from dollars. Nonetheless, neither Barb nor Carla were convinced that the idea made sense. ... What did make sense to Barb and Carla was to apply lessons they had learned from their own prior "real life" experiences to help them make decisions in case of the "summer jobs" problem. Therefore, the team didn't pursue Alan's suggestion. Instead, Alan's suggestion was used as a (transitional) way of thinking which led to a better idea which Barb suggested that would take into account *both* time and money.

Barb: Hey, that's a good idea! We could figure out dollars-per-hour. ... I did that for my jobs last summer.

Interpretation #9: Focusing on Dollars-per-hour

Barb wasn't really paying close attention to Alan's idea. The new ideas that she heard were to think about the situation in the same way that she thought about her own past jobs. That is, both Alan and Barb were using past "real life" experience to make sense of the current problem. Therefore, Barb thought in terms of dollars-per-hour.

For the remaining minutes of the class, Alan, Barb, and Carla went back to the original data tables and started calculating dollars-per-hour. ... As class ended, they decided that, to prepare for the next day's class, each student should bring a graph showing dollars-per-hour for the workers. Then they planned to use these graphs to make final decisions about who to hire. The graphs on the following page show what each student brought to class the next day.

Interpretation #10a: Alan's dollars-per-hour graph based on sums for the whole summer

First, Alan calculated the total amount of money that each worker earned for the whole summer. Then, he calculated how much time they worked altogether. Finally, for each worker, he divided total dollars by total time.

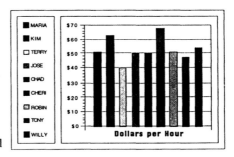

Interpretation #10b: Barb's dollar-per-hour graph based on sums for each month

First, Barb calculated the total amount of money that each worker earned for each month. Then, she calculated how much time they worked each month. Finally, for each month, she divided dollars earned by time worked.

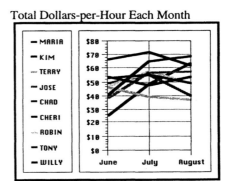

Total Dollars-per-Hour Each Month

Interpretation #10c: Carla's graph showing the average dollars-per-hour each month (where the average is taken across busy, steady, and slow periods).

	June			July			August		
	Busy	Steady	Slow	Busy	Steady	Slow	Busy	Steady	Slow
Maria	$55.20	$52.00	$50.22	$69.90	$54.14	$47.71	$63.04	$51.70	$41.77
Kim	$86.18	$39.73	$26.19	$86.21	$50.80	$30.77	$90.00	$59.57	$30.30
Terry	$87.25	$39.24	$19.59	$69.45	$32.16	$20.93	$54.46	$39.32	$20.04
Jose	$64.77	$38.95	$22.50	$79.20	$53.81	$32.07	$82.82	$65.44	$37.72
Chad	$64.82	$45.08		$68.81	$43.94	$20.30	$64.10	$47.08	$19.78
Cheri	$85.77	$61.78	$47.83	$88.72	$63.97	$35.54	$82.63	$66.42	$34.97
Robin	$85.02	$39.13	$22.59	$66.72	$38.19	$25.00	$66.36	$40.12	$15.82
Tony	$73.33	$56.44	$37.12	$81.00	$51.87	$51.18	$82.00	$52.00	$29.98
Willy		$41.67	$14.22	$80.87	$43.83	$19.69	$81.22	$56.95	$21.08

Interpretation #10c, continued

Note: Carla got some help from her brother, who apparently suggested the idea of an average. First, Carla calculated the dollar-per-hour for each cell in the matrix shown. Then, for each month, she calculated the average of the rates for the busy, steady, and slow periods. This procedure assumes (incorrectly) that the students intended to treat busy, steady, and slow periods as being equally important!

Average Dollars-per-Hour Each Month
(Across Busy, Steady, & Slow Periods)

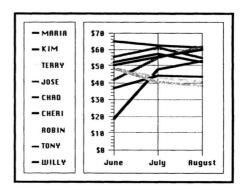

For approximately the first twenty minutes of the second class, Alan, Barb, and Carla showed one another their rate-per-hour graphs, and they explained how the graphs were made. Then, for each graph, the team as a whole worked to try to decide which workers should fall into the categories: full time, part time, and don't hire. For Alan's list, the ranking was easy to read directly from the graph that he had drawn. But, for Barb's graph and for Carla's graph, it was not as obvious to determine which workers ranked first, second, third, and so on. Therefore, for both of these graphs, the teams used telescoping methods of decision making. That is, first-round (tentative) decisions were based on performances in August alone. Then, to make decisions about difficult cases, information was used from July (or from June). The results are shown below.

Interpretation #11: Three different lists were generated that ranked workers from lowest to highest based on the dollar-per-hour graphs that the students had produced.

	Alan's List	Barb's List	Carla's List
	Cheri	Kim	Cheri
FULL TIME	Kim	Cheri	Jose
	Willy	Willy	Kim
	Maria	Jose	Tony
PART TIME	Robin	Chad	Maria
	Chad	Robin	Willy
	Jose	Maria	Chad
DON'T HIRE	Tony	Tony	Robin
	Terry	Terry	Terry

Because the preceding three lists were somewhat different, Alan, Barb, and Carla tried to make a new list (which they called their "agreement list") showing points of agreement among the three lists. While the students are discussing the possibilities, Carla comes up with a new idea.

Carla: Look, on my list, Cheri, Jose, and Kim all got A's. ... Tony, Robin, and Willy got B's. And, Chad, Robin, and Terry got C's. ... What did they get on your lists?
Alan: What do you mean?
Carla: Give me your list, I'll show you. ... {pause} ... See. Cheri got an A, and so did Kim and Willy.
Barb: What are you guys doing?
Carla: Here watch.

For approximately the next five minutes, Carla asks the other two students to give her information to fill in the "grading scale" shown on the following page.

Interpretation #12: Generating a "grading scheme."

For each list, a "grading scheme" is imposed that is similar to those used for tests in class. The scores are then combined (treating each of the rankings as if they were independent ratings).

	Alan's List	Barb's List	Carla's List	Combined
Cheri	A	A	A	A
Kim	A	A	A	A
Willy	A	A	B	A-
Jose	C	B	A	B
Robin	B	B	C	B-
Chad	B	B	C	B-
Maria	B	C	B	B-
Tony	C	C	B	C+
Terry	C	C	C	C

Alan: So, it looks like the full time people should be Cheri, and Kim, and Willy. ... And part time should be Jose, and ... uh oh! Who should we pick next? Maria, Robin, or Chad?
Barb: Yeah. Tony and Terry are out.
Alan: These other guys are pretty close. ... It's not fair to just pick one.
Carla: Maybe one of these guys really needs a job. I'd think we should hire guys who really need a job. Maybe Willy doesn't really need a job. Maybe Jose really needs one.
Alan: Some of these guys probably didn't get to work at the good times. {pause}

Barb: Let's make more graphs like these {pointing to her rate-per-hour graphs in interpretation #10C} for the slow times, and the steady times, and the fast times.

More than 12 minutes pass while Alan, Barb, and Carla worked together to make graphs comparing dollars-per hour for busy, steady, and slow periods.

Interpretation #13: A telescoping series of rules.

First round decisions are based on interpretation #12. Then, second round decisions are made by comparing dollars-per-hour for busy, steady, and slow periods. (The graph for busy periods is shown, similar graphs were made for steady and slow periods.)

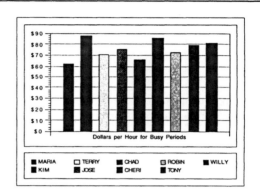

Barb: {looking at the preceding graphs} I don't think this helps much.
Carla: {looking at the preceding graphs} So, which one should we hire? Maria, Robin, or Chad?
Alan: Look, Maria's only best during slow times. But, we don't really care about slow times. We're only going to hire part time people when things are happening ... fast times. {pause}
Barb: Wait a minute. Maria's not so bad. Look, um, she's better that Robin during steady times. ... and Chad too.

Approximately eight minutes pass in which Alan, Barb, and Carla looked back over the graphs that they brought to class, and the work they did earlier in the period. In these discussions, they offer "stories" that might possibly explain patterns in the dollars-per-hour for various workers. In the end, they reached an agreement on the following points:

1) slow periods should not be treated as being very important, because (a) most of the money would be made during busy or steady periods, and (b) part time workers would not be hired during slow periods.

2) performance in August (and, to a lesser extent, July) should be treated as being most important, because (a) it took into account learning and improvement, and (b) it was the most recent indicator of worker capabilities.

Carla: We've got to write up our report. ... What should we do?
Barb: I think we should make another graph like the one I made before {i.e. Interpretation #10b} ... only this time leave out slow times.
Carla: OK, you do that. ... I'll get the poster board and stuff.

For the remainder of the class period, Alan, Barb, and Carla worked together to produce a large poster like the one shown on the following page.

Interpretation #14: A telescoping series of rules based on dollar-per-hour trends for (only) busy and steady periods.

Dear Maya,
We think you should hire Kim and Cheri and Jose for full time, and we think you should hire Willy and Chad and Tony for part time. Look at this graph to see why these people are best.
The graph is only about busy times and steady times. You don't make much money during slow times, and you won't hire people for slow times.
Some workers got better at the end of the summer. But, some didn't get better. So, August is most important, and July is also important. July is when you make the most money.

Alan, Barb, and Carla

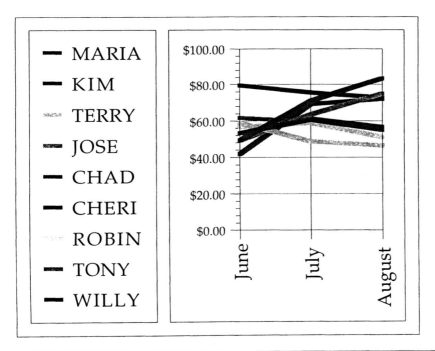

IMPLICATIONS

The interpretations provided for the responses given by Alan, Barb, and Carla were intended to help the reader gain insight into the various "models" and "modeling cycles" that were used during the solution process. Notice that the students cycled through models that began with informal intuitions (performing simple computations without any interpretation or mathematization of the data) and proceeded toward more formal systems (looking at trends, averages, and graphical analysis and representations of data). Furthermore, the students began to go beyond thinking *with* conceptual models to also thinking *about* them. That is, they analyzed the underlying assumptions, strengths, and weaknesses associated with each model. For example, they could have stopped after calculating "hours worked" as in interpretation #2, but Carla pointed out that they should look at money earned before making a decision. Or, they could have decided who to hire based upon total number of dollars earned (as in interpretation #3), but Barb pointed out that "some guys got to work a lot more than others." Each revision made during the solution process involved related, but qualitatively distinct models, and these were needed to produce the increasingly useful solutions to this problem. If the students were not given sufficient time, resources, and opportunities to cycle through the different models, powerful ideas may never have been developed and applied.

It is useful to note that the teacher was not the one who prompted the students to use the different conceptual models. In fact, she did not intercede or interfere with the flow of events or ideas. She did however, create an environment in which students could discuss, defend, and justify their own solutions. In addition, she provided the students with access to realistic tools and resources such as calculators and computers.

What prompted the students to develop and use the powerful conceptual models discussed in the previous section? Consider the characteristics of "model-eliciting" activities stated previously. First, the problem focused on a "real" issue which required a decision that needed to be addressed (not just a school-math question related to it). The students knew who was asking for the information (Maya), why she was asking (to help her decide who to rehire), and the criteria that would ultimately influence the quality of the response. Next, the students needed to justify and explain their decisions by describing underlying assumptions and conditions, and they had to analyze and assess alternative conclusions, explanations and interpretations generated by themselves and each other. Third, the statement of the problem required them to explicitly reveal how they were thinking about the situation (the givens, goals, possible solution paths, etc.—recall that Maya asked them to provide a clear explanation to help her decide whether the method chosen for evaluating the vendors is a good one for her to use). Fourth, the students had to make judgments for themselves about issues such as whether (or in which directions) current solutions needed to improve. Fifth, the problem prompted the students to consider models which ultimately can be used to generate answers to a whole class of questions in a whole range of situations (models for dealing with trends, averages, data analysis, etc.). This is in sharp contrast to textbook problems which generally ask students to produce nothing more than a specific response to a particular question, and not produce a model at all. Last, the problem situation was designed to be as simple as possible while still creating the need for a significant model. In total, these characteristics pushed students to construct, manipulate, extend, and refine powerful mathematical models.

The interpretation boxes were intended to highlight the statistical uses that emerged as the students solved the problem activity. These included: multiple views of data, both graphical and tabular; measures of central tendency; analyses of trends; and procedures for combining data. Since mathematical knowledge and abilities develop along a number of dimensions such as from concrete to abstract, from specific to general, from global/undifferentiated to refined, or from intuitions to formalizations, it is important for teachers to use model-eliciting activities at the beginning of instruction to gather valuable assessment information. As students work on these activities, they not only reveal the concrete/intuitive/informal understandings (and misunderstandings) that they have, but they can also extend, refine, or integrate these ideas to develop new levels of understandings. One of the points to be made with regard to assessment, is that when teachers observe students solving model-eliciting problem activities, the information that they get is similar to the kind that might have resulted from one-to-one clinical interviews. Therefore, teachers learn about students' strengths and weaknesses so that: (1) they can avoid re-teaching ideas that students already know; (2) they can focus on the key issues that students do not understand; and (3) they can use existing understandings and capabilities as foundations for new knowledge and abilities. Model-enhancing activities can be used as part of a student's portfolio (see Chapter 13). Note however, they differ quite significantly from multiple-choice methods (see Chapter 16).

In sum, this chapter was intended to show that when students are provided with opportunities to solve model-eliciting activities in which they can assess and monitor their own work using realistic tools, and when more options are available concerning modes of responses and solution paths, students can construct, modify, and refine powerful conceptual models for dealing with data analysis and statistics. We believe that model-eliciting tasks are important in statistics education because they offer a window into statistical reasoning processes or conceptual structures in statistics education that otherwise may be difficult or impossible to assess using traditional methods.

Chapter 7

Simple Approaches to Assessing Underlying Understanding of Statistical Concepts

Anthony E. Kelly, Finbarr Sloane and Andrea Whittaker

Purpose

Statistics should be introduced with clear linkages to the mathematics that students already understand and within contexts that students find meaningful. Otherwise, students may learn statistics in a rote fashion or apply statistics in a merely instrumental fashion and draw erroneous conclusions from data. In this chapter we present two examples of the use of simple assessment techniques that uncovered students' poor understanding of statistical concepts.

INTRODUCTION

Many people reason in ways that contradict accepted statistical models (Tversky & Kahneman, 1971; Fong, Krantz, & Nisbett, 1986; Konold, 1991a,b). Among 11- to 16- year-olds, statistical reasoning and routines are often viewed as arbitrary and inaccessible (Green, 1983). Problems in learning statistics persist even among researchers in the behavioral sciences (e.g., Greer & Semrau, 1984) and in medicine (e.g., Clayden & Croft, 1989). Medical researchers frequently confuse: histogram and barchart, correlation, standard deviation and variance (Clayden & Croft, 1989).

Completing a course in statistics does not inevitably lead to statistical insight. In a number of studies, students in statistics courses were found to: (a) describe rather than justify their statistical solutions (Allwood & Montgomery, 1982); (b) fail to establish a conceptual base for their solution strategies (Allwood & Montgomery, 1981); and, (c) when faced with errors in their statistical solutions misjudge their errors as correct (Montgomery & Allwood, 1978a, 1978b), or ignore their incorrect substeps when accounting for their solutions (Allwood & Montgomery, 1981).

Many students learn statistics as a set of rules without always learning the meaningful contexts in which they should be applied. Skemp (1979) describes mere rule-based learning as "instrumental understanding" consisting of recognizing a task for which one knows a particular rule. What we wish to strive for in our instruction is learning with meaning, what Skemp calls "relational understanding." In relational understanding, one relates a task to an appropriate schema or model; one does not blindly apply rules.

The difficulty for teachers of statistics is to recognise those cases in which instrumental understanding is being passed off as relational understanding. One way to guard against students' learning without understanding is to adopt assessment techniques that allow the teacher some insight into students' thinking. These techniques should be relatively easy to use, and economical of the teacher's time. This chapter will illustrate with two examples how assessment approaches that focus only on computational aspects of statistics may miss misunderstandings that can be exposed by judicious choice of tasks, and interview questions. In the first example, students' approaches to describing simple data were examined. In the second example, students' understanding of the analysis of variance technique were explored.

DESCRIBING SIMPLE DATA

Twenty-five graduate students took a non-compulsory, introductory-level statistics class. In the course, they were instructed in the use and value of plots from an exploratory data analysis (EDA) perspective, and were introduced to the SAS statistics package.

Assessing shape is an important first step in data description, whether the data summary is to be graphic, numeric, or verbal (see Moore, 1990). Data distribution can be assessed by drawing a bar chart (for a discretely measured variable), a histogram (for a continuously measured variable), or a stem-and-leaf plot. The students were asked to respond to the following question:

> The 'simplest' form of statistics is to summarize a set of univariate data. Summarize the following data in whatever way you think is appropriate. The data refer to the heights (in meters) of 20 women who are being investigated for a medical condition:
>
> 1.52 1.60 1.57 1.60 1.75 1.63 1.55 1.63 1.55
> 1.65 1.55 1.65 1.60 1.68 2.50 1.52 1.65 1.65

To test for mere instrumental understanding of statistical techniques, there are two deliberate misprints in this problem: (a) there are only 18 measures (not 20 as claimed); and (b) the observation of an 8-foot woman (2.50 meters) is almost certainly an error. The first misprint is designed to capture mechanical applications of procedure. The second is designed to capture mindless acceptance of data as "given." A central danger in statistics education is that we neglect to tell our students that statistics deals not with numbers, but with numbers in context (Moore, 1990). For the latter error, a student could be expected to either note the data-entry error and provide a biased estimate of the mean, change it to 1.50 meters or other reasonable entry, and note the correction, or omit it and calculate a mean with the remaining 17 measures. In all cases some justificatory statement arguing that the 2.50 meters is a miskeying on the grounds that 8-foot tall females are not likely to exist would be expected.

Categories of responses

We have categorized the student responses in four ways: mindless reliance on statistics packages, generating a mean value without plotting the data, plotting the data and generating a mean value uninformed by the plot, and plotting the data and linking the choice of appropriate statistic to the distribution of the data.

Mindless reliance on statistics packages. One student responded, "I would enter these [data] into SAS and use the PROC UNIVARIATE function. This would give us all the info we needed and more: including mean, mode, median, standard deviation, variance, range, etc." The student failed to link the choice of appropriate statistic to any underlying distributional assumptions. Note that, ironically, the statistics package would generate the correct sample size.

Generating a mean value without plotting the data. The concept of the mean may appear so simple and unambiguous that adult students should not have any difficulties in understanding or using it. The mean can be seen frequently in everyday life (for example, in professional and college sports, in the construction of high school, college and graduate school GPA's). Most data reported in professional journals are means and inferential statistics that deal almost exclusively with means and mean differences, yet the mean causes problems for many learners (Pollatsek, Lima and Well, 1981). In this study, ten students provided a mean value without the use of a graphical method. Of these, four students trusted the claim that there were 20 measures in the study. These four students, consequently, reported a *mean value that lay outside the range of the data*— impossible for a correctly computed mean.

Plotting the data and generating a mean value uninformed by the plot. Students who plotted the data used a variety of graphical methods. These included frequency tables, stem-and-leaf plots and histograms. Six students constructed frequency tables of the sort displayed below:

Table 1. Example Student Frequency Table

Measures	Frequencies
1.52	ll
1.55	lll
1.57	l
1.60	lll
1.63	ll
1.65	llll
1.68	l
1.75	l
2.50	l

As the intervals between measures are uneven, tables of this type do not help the students see the effect of the outlying value. Moreover, two of the six students summed the number of frequencies to 20—the number provided in the problem statement—although their tallies indicate that there are 18 measures. The two students who constructed histograms failed to hold the areas, or the intervals constant across the x-axis. This resulted in their inability to see the measure 2.50 meters as being an outlier.

Plotting the data and linking the choice of statistic to distributional assumptions. Of the remaining eight students who constructed stem-and-leaf plots, five reported a biased mean value. One of these students, when probed, noted that, "graphing data is the right thing to do. You graph your data first, and then get your statistics." Only three of these eight noted 2.50 as an outlier when generating a mean value. These same students also realized that only 18 observations were available to them and moreover they based their choice of statistic on the distributions they had constructed. Student #17 constructed the following display, and noted, "Clearly woman 15 is very tall and should be considered an outlying value."

Table 2. Stem-and-leaf Display (student #17)

```
2.5  0
2.4
2.3
2.2
2.1
2.0
1.9
1.8
1.7  5
1.6  0033550855
1.5  275552
```

From the interview data it became apparent that all of the students had at least an instrumental level of understanding. Most of the students described here were able to employ a rule (or procedure), but many saw little or no connection between the generated answers and their meaningfulness, or the meaningfulness of the original data themselves.

ANALYSIS OF VARIANCE

Graduate students who had scored either a "B" or better in a course on analysis of variance were interviewed. To earn a grade of "B," the students had to successfully complete periodic assignments in which statistics packages were used to analyze prepared data sets, and to complete a final exam involving computations. In this study, students were able to mask their conceptual confusion during the course when they worked on prepared data sets (on paper and on the computer), and when they took the final written exam. Here we characterize some of the misconceptions of students who had "successfully passed" the course.

The students were asked, individually, to: (a) talk about the new statistical concepts in their own words; (b) to apply the statistical concepts to data sets that were *unlike* the ones that they had met in the course; and (c) explain how the statistical concepts would help them interpret the results of experimental studies.

Eight of eleven students interviewed had memorized fragments of statistical knowledge that were sufficient to "earn" good grades on written examinations, but that did not form a solid basis for advanced statistical knowledge. Without concerning ourselves with the details of analysis of variance, we can say that the students' responses could be characterized in a number of distinct ways. We again noted evidence of merely instrumental understanding of statistics.

Four students knew the appropriate goal of the statistical technique, but were unable to discuss the technique with understanding (e.g., "ANOVA is for looking at means, so I would use ANOVA."). In response to questions, they parroted poorly remembered statistical fragments or labels (e.g., "I would get the 'mean square within' and the 'mean square between'"). They were like tourists with a destination who could not locate the foreign language phrase book.

Two other students were like tourists who had mastered only the grammar of the language. They replied to conceptual questions with only mathematical statements or descriptions of computational routines. They were unable to connect the manipulations to word problems (e.g., "To solve this problem I would divide the 'mean square within' by . . .")

Two further students evidenced the opposite problem: they could interpret the terms in the statistical model using everyday vocabulary, but could not link that understanding to the underlying mathematics (e.g., "Well, you have a measure of the signal in the data and a measure of the noise in the data, and you place them in a ratio, but I am not sure which numbers to use"). They were like people who can drive, but who are powerless when the car malfunctions. When asked to "look under the hood" of the statistical techniques, they were unable to explain their higher-level reasoning in terms of the underlying equations.

One insight into instrumental understanding was the following. The probability of a Type I error—claiming there is a difference among groups when the null hypothesis is true— is conventionally set at $p < .05$. One of the students, who could apply this rule correctly on written assignments, when interviewed, argued that the probability of a Type I error should be large as possible because a p valueless than .05, was *such a small number* that it implied that the differences among the groups must be *insignificant:* the reverse, of course, being the case.

Clearly, for teaching and assessment purposes, a distinction must be drawn between instrumental and relational learning, between computational skill and statistical expertise. In both of these studies, computational assignments provided the students with the opportunity to mask poor relational understanding with (somewhat) effective instrumental understanding. When "traps" were set in test items, and the students were interviewed, their misconceptions came to light.

IMPLICATIONS FOR ASSESSMENT

Realize that the students must construct their own meaning for new concepts, but sometimes resign themselves to learning material mindlessly. If students do not make sense of what we teach both in terms of their own day-to-day language, and in terms of the language of mathematics, they may compartmentalize it, and respond to tasks instrumentally or robotically. As we have seen, both simple and advanced statistical tests can be calculated mindlessly. To guard against this, we must (in our teaching and assessment) explicitly link the modeling language of mathematics to the conceptual model we are developing (see, for example, Lovie, 1978 and Lovie & Lovie, 1976). If we do not link more familiar concepts with statistical routines, we risk teaching (and later assessing) an easily forgotten cryptology such as "PROC UNIVARIATE" or "mean square between divided by mean square within" whose symbolic mathematical explication is an even more alchemical incantation: "Sigma X sub i sub j..."

Don't just talk, listen. It does not profit our students to "talk at them" a subject matter that, if instrumentally learned, will have no value beyond the "final" for the course. Worse still, students may leave the course with an unjustified sense of mastery. We must design tasks that require students to be mindful, tasks that eschew the sanitized and "perfect" items of statistics textbooks. Then we must listen to their responses. Do not use the form: Here is the problem, calculate an answer. This approach failed in the written examinations described above. Rather, we should adopt the role of the naive listener who seeks continuous clarification: "You told me that I needed to look at the data. What does that imply here? I understand that a mean is a 'measure of central tendency.' What would it look like in this data set? Is it the same thing as a mode? Why not? Is it better than a mode? Can there be more than one mean in a data set? Explain your procedure to me in your own words," and so forth. Or more generally, we can ask, "Does the data make sense to you? Does your answer make sense to you?"

Listen not just to the content of the response; listen also for the affect and motivation of the student. Does the student sound confident when giving the response? As you listen, does the image come to mind of a traveler striding confidently along boulevards or a tourist fumbling sheepishly through an English-Statistics/Statistics-English dictionary? If it is the latter, remember that assessment is best when it is in the service of the student. Do not embarrass the lost traveler. Turn the student's bewilderment into an opportunity to revisit a concept that may be a concern for more than this one student.

Don't be misled by correct-sounding answers or flawless technique. As we listen to our students, we must guard against the error of assuming that, just because we hear the correct-sounding terminology, the students understand what they are talking about. Some students learn quickly how to manipulate mathematical symbols so that they get an "answer" (see Dallal, 1990). We must constantly look beyond the information given to us. Some high school students were asked by the first author to give an example of a "variable." They answered, "x." When asked what "x" was, they replied, correctly, but circularly, "A variable"! When asked for a *different* example of a variable, they replied, "y"!

Statistics is an interpretive science. It involves defensible descriptions, and defensible inferences, about samples and populations. Statistics is integrally related to models in the world and critical problem-solving skills. Statistical thinking involves understanding the characteristics of a problem well enough to be able to select and apply the appropriate tool to answer the problem. It is not enough to be able to "do" the routine (either by hand or on the computer); one must know why one has chosen it, what its application tells one, and what limitations one must place upon its conclusions.

Chapter 8

Assessing Students' Connected Understanding of Statistical Relationships

Candace Schau and Nancy Mattern

Purpose

We believe that connected understanding among concepts is necessary for successful statistical reasoning and problem solving. Two of our major instructional goals in teaching statistics at any level are to assist students in gaining connected understanding and to assess their understanding. In this chapter, we will explore the following questions:

- Why is connected understanding important in statistics education?
- What models of connected understanding are useful in thinking about statistics learning?
- How can connected understanding be represented visually?
- What approaches exist for assessing connected understanding?

INTRODUCTION

We, and many other statistics instructors, routinely observe a critical weakness in post-secondary students who have taken applied statistics courses: they lack understanding of the connections among the important, functional concepts in the discipline. Without understanding these connections, students cannot effectively and efficiently engage in statistical reasoning and problem-solving. They remain novices. They have "isolated" knowledge about various concepts; for example, they may be able to calculate a standard deviation and a standard error. However, they do not understand how these concepts are related (and distinguished) and so make application mistakes such as using one concept when they should have used the other. Some students recognize their lack of connected understanding and will say things like "I can solve a problem using the t-test when I know I'm supposed to. But otherwise I don't have a clue."

There is a growing body of research with findings that attest to the statistical and probabilistic misconceptions held by students of all ages, as well as by adults. Garfield and Ahlgren (1988) cite study after study concluding that students cannot use statistical reasoning effectively in probabilistic situations nor can they solve statistical problems, even after exposure to instruction in statistics. We contend that these difficulties are due to students' lack of connected understanding of concepts.

The current work on national standards and goals in our country supports our belief in the importance of conceptual connections. The *Curriculum and Evaluation Standards for School Mathematics* (National Council of Teachers of Mathematics, 1989) explicitly contains a standard for each set of grades called "Connections" (standard #4). This standard for Grades K-4 includes five desired outcomes such as "link ... knowledge" and "recognize relationships" (p. 32). For Grades 5-8, the five desired outcomes include, for example, "see mathematics as an integrated whole" (p. 84). For Grades 9-12, the four outcomes include, for example, "use and value the connections among mathematical topics" and "use and value the connections between mathematics and other disciplines" (p. 148).

To determine the extent to which both formative and summative instructional connection goals are met, instructors need assessment formats that measure connected understanding. This chapter describes our thinking about this kind of assessment. It is based on our work studying post-secondary students' connected understanding of statistics, as well as middle-school and post-secondary students' connected understanding of science.

MODELS OF CONNECTED UNDERSTANDING

There are many ways to think about learning. One fruitful way for us as teachers and researchers has been to think about the mental networks students form in the process of learning and use in applying their learning. The schema is an important concept in many theories of mental networks. Schemas and connected groups of schemas often are called cognitive structures, cognitive networks, or structural knowledge (Skemp, 1987).

A schema is a mental storage mechanism that is structured as a network of knowledge (Marshall, 1995). A schema results from repeated exposure to problem-solving situations that have features in common. The learner forms a schema by abstracting the most relevant of these features and either assimilating these features into existing schemas or creating new schemas. One critical defining feature of a schema is the presence of connections; in order to function, the components within a schema must be interconnected. As students (who are "novices") become more expert, their schemas gain components and these components and the schemas themselves become more interconnected; their cognitive structures begin to resemble those of their instructors (who are "experts"). Expertise is not merely knowing "a lot"; it is having a rich, accurate, and relevant set of interconnected schema and schema components (Marshall, 1995; Skemp, 1987). According to this model, then, students learn statistics through 1) assimilating (connecting) new information into their cognitive networks, 2) forming new connections among knowledge that already exists in their networks, 3) reorganizing (accommodating) their connected schema to match incoming information, and 4) eliminating incorrect concepts and connections.

Learners can (and often do) develop inaccurate schemas. Inaccurate schemas are most often formed when one or both of two conditions exist. The first condition reflects the state of the learner; the second concerns the problem-solving tasks learners experience (Skemp, 1987). First, all but the most basic schemas are built on existing schemas. Some students lack these prerequisite schemas. Others may have them, but they may be inaccurate; that is, some learners possess misconceptions when they enter statistics classes.

Second, problem-solving tasks may interfere with accurate schema formation in one of three common ways. Students may not experience enough tasks to be able to abstract the needed relevant common features. The tasks may lack these features. The tasks may have many features in common, and students may not be able to determine which are relevant and which are not.

Whether learners lack prerequisite schemas or problem-solving tasks are inadequate, resulting statistical schemas aren't developed at all, are incomplete, or are inaccurate. Both sets of situations require instructional planning and intervention. Statistics instructors need to identify and address important misconceptions before moving on to more complex knowledge. They also need to carefully develop (or select and evaluate) the problem-solving situations they use in instruction.

Schemas develop because they are useful in problem-solving. When faced with a task, problem-solvers first use their existing schemas to create a mental model of the problem. They then use schemas to plan an approach to solving the problem, to execute that approach, and to evaluate the outcome. The problem-solvers stop if the process was successful, try again until it is, or quit trying. To begin this process successfully, problem-solvers must recognize the critical features of the problem and match them to existing schemas; if students fail to form a mental model at this stage or if they form an incorrect mental model, the entire problem-solving process fails (Marshall, 1995; Skemp, 1987).

Statistics students can attempt to form mental models in at least two ways. First, they can try to abstract the important components of the problem and match these to their existing schemas. For the student to be successful, these schemas must be relevant and accurate. This approach is flexible and adaptable to many different problem-solving situations and tasks. Alternatively, if students do not possess relevant and accurate schemas or if they do not invoke them, they can attempt to form their mental models through trying to recall the appropriate formula. This approach becomes increasingly harder as more and more formulas are encountered in learning. Initially, it may be easier to memorize the formula needed to solve a particular problem than to form a schema, but eventually the sheer number of formulas and different problems becomes overwhelming for students' memories (Skemp, 1987). In addition, there are many problems (perhaps the most important ones) that cannot be directly solved by "plugging" into a single formula and "chugging" out the answer, either by hand or on a computer.

MAP REPRESENTATIONS OF CONNECTED UNDERSTANDING

Understanding in statistics, then, requires the existence and use of relevant and accurate schema that, by their nature, are organized in a mental structure. This organization can be represented as a visual-spatial network (or map) of connected concepts. Perhaps the most widely known, and inclusive, of map formats is the concept map (Novak & Gowin, 1984). A concept map includes the concepts (referred to as nodes and often represented visually by ovals or rectangles) and the connections (referred to as links and often represented with arrows) that relate them. In this representation, the basic unit is a pair of connected concepts called a proposition and consisting of two nodes connected by a link. Like schema networks, maps can, but do not need to, include hierarchical structures. See Holley and Dansereau (1984); Jonassen, Beissner, and Yacci (1993); and Novak and Gowin (1984).

Good concept maps include enough important concepts to represent the targeted subject area, and these concepts are clearly linked with important relationships. Figure 1, for example, presents our concept map representation of one possible hierarchical network that is relevant to us when we teach introductory statistics. In Figure 1, the proposition "statistics can be inferential" consists of the concepts of "statistics" and "inferential" connected by the link "can be." Propositions often are grouped into neighborhoods (clusters of propositions that are more densely related to each other than to other propositions). Figure 1 contains two neighborhoods, one containing propositions related to descriptive statistics and the other propositions related to inferential statistics. The *inferential* neighborhood contains two subneighborhoods, one related to "Variables that are Continuous" and the other to "Variables that are Discrete."

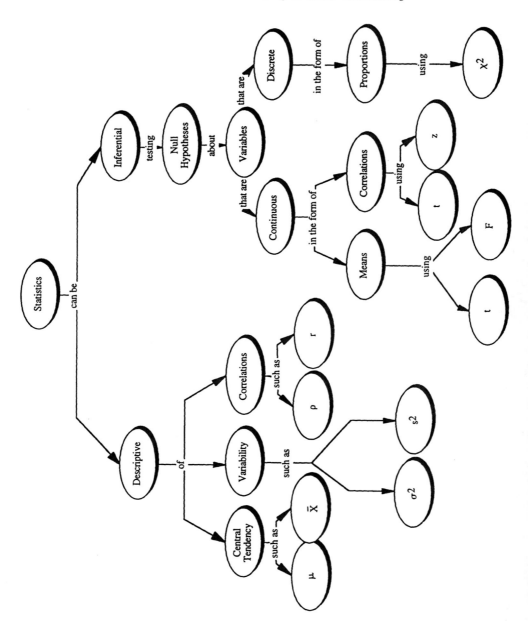

Figure 1. "Expert" (instructor-made) concept map representing aspects of one organization for an introductory statistics course

Figure 2 is a hierarchical map we created to represent the process of hypothesis testing of means. It contains two neighborhoods; the neighborhood under "Discrete independent variable(s)" contains two subneighborhoods. Neighborhoods may be cross-linked with each other; the "tested using" link connects neighborhoods 1 and 2 in this figure.

We construct "expert" maps like those in Figures 1 and 2 in several steps. We first identify the purpose that the map is to serve (an extended version of Figure 1, for example, was designed as a global map of course content). Second, we select the concept(s) that will serve as the foci for the map (in the case of Figure 1, the central concept is "Statistics"; second-level concepts are "Descriptive" and "Inferential"). Third, we identify the important concepts that are related to the central and second-level concepts and how these concepts are related. Fourth, we roughly sketch the map. Fifth, we draw the map using a software package called Inspiration (1994) or by hand. Sixth, we discuss our map and revise it, often by eliminating concepts, working on links, and rearranging propositions and neighborhoods.

Concept maps and other mapping techniques are used for at least three instructional purposes. First, they are used as an instructional planning tool. Teachers can use mapping in designing content coverage materials (such as handouts and notes), sequencing instructional delivery, and determining delivery strategies (such as activities and presentations). Creating these maps forces instructors to identify important connected concepts. It is impossible to draw maps without making explicit your implicit mental characterizations. In our experience, what we do not understand becomes painfully clear as we try to draw our maps. Second, maps are a learning tool (Harnisch, Sato, Zheng, Yamagi, & Connell, 1994). They may be prepared by a teacher and then shared with students with the intent of portraying a representation of the instructor's connected understanding of the material. Instructors also can assign map creation to students to encourage connected understanding and schema-building. There is a growing body of research indicating that both of these uses assist many students in learning. Third, these mapping formats can be and are used for both formative and summative assessment, either through map creation or map completion by students. Concept maps are the mapping format most widely used for assessment.

Our theory of how students learn should drive our development and use of appropriate assessments (Marshall, 1995). We believe that understanding requires connected concepts. If connected understanding is an important instructional goal, we must be able to assess it. In this chapter, we emphasize the use of concept maps for assessing connected understanding, contrasting the maps with assessments consisting of traditional item types.

USING TRADITIONAL ITEM TYPES TO ASSESS CONNECTED UNDERSTANDING

Two traditional item types—multiple choice items and word problems—have a long history of use for assessing achievement in statistics and mathematics. The common approach to the use of traditional assessment items includes the assumption (usually implicit) that understanding is an accumulation of bits of isolated information (Marshall, 1995). A test often consists of a collection of items measuring primarily recall.

Both of these item types can be used to measure aspects of connected understanding. Even when they are written to assess cognitive structures, they yield limited information: more *information is obtained from analyzing errors than by analyzing correct answers.*

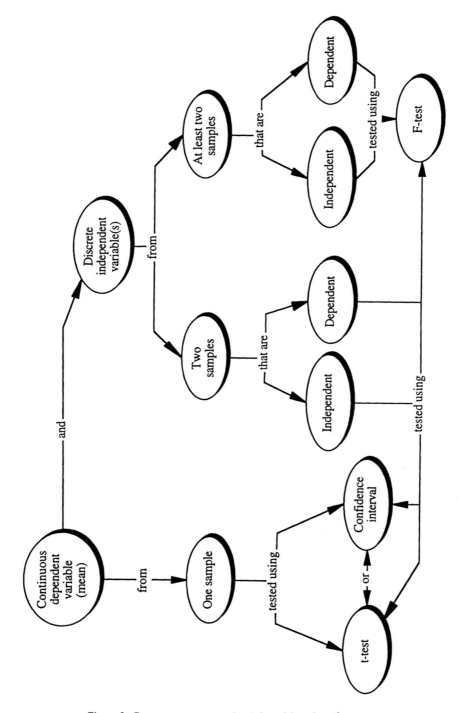

Figure 2. Concept map representing inferential testing of means

Multiple-choice items

Many multiple-choice items are written to assess the quantity of facts students can recall. The following item, for example, assesses recall of a symbol.

The sample standard deviation is symbolized by
   a. σ
   b. σ^2
   c. s
   d. s^2
   e. SS

The distractors in the above item contain symbols that are correct for other measures of variability. Students will have seen all of these symbols previously in class and in their textbook.

It is likely that most students will attempt to answer this item using fact recognition. However, students who understand that Greek letters are associated with population parameters and Roman letters with sample statistics could use that understanding to eliminate distractors "a" and "b." Even on some recall items, students with connected understanding can have an advantage (if they think to use that understanding).

We can write multiple-choice questions that require connected understanding in order to arrive at the correct answer (unless, of course, a student is lucky enough to guess the correct answer). These items include distractors that lead students with incorrectly connected concepts to select a wrong answer. They are difficult and time-consuming to write. In addition, each item can assess only a small area of the students' networks.

For example, the following item assesses aspects of students' connected understanding of standard deviation.

A group of 30 introductory statistics students took a 25-item test. The mean and standard deviation were computed; the standard deviation was 0. You know that:
   a. about half of the scores were above the mean.
   b. the test was so hard that everyone missed all items.
   c. a math error was made in computing the standard deviation.
   d. everyone correctly answered the same number of items.
   e. the mean, median, and mode from these scores probably differ.

The correct answer is "d." Students who connect standard deviation to the normal curve and z scores may choose distractor "a." Students who confuse the mean and standard deviation may choose option "b." If students have not seen an example or application in which the standard deviation equaled 0, they may decide that this value is impossible to obtain and choose option "c." Students who remember that values for the three measures of central tendency usually have differed in their examples may choose option "e."

We used this item on the first test in our introductory graduate-level applied statistics course. Students were allowed to use extensive notes but not their textbooks. About 75% of our students answered this item correctly. It discriminated well between higher and lower achievers; that is, many more students with higher total test scores answered this item correctly than did students with lower total test scores.

Word problems

Word problems may be the most frequently used format for assessment in K-12 mathematics instructions as well as in small post-secondary mathematics and statistics courses. Typically, the student is presented with a set of raw or summary data and asked to do something with that information. We used the following word problem on the first test given in the course we described above. Students had worked similar problems, but none was identical to the test item.

> Telia scored at the 90th percentile in grade 8 on a standardized test given at the beginning of the school year. The mean grade-equivalent (GE) score of the 8th-grade norm group on this test is 8.0 and the standard deviation is 2.0 GE. What was Telia's GE score?

The solution to this problem involves transforming Telia's percentile score to a z score using a unit normal (z) distribution table, then solving for Telia's GE score using the z score formula. Again, our students had access to notes.

Our students tried to build three common mental models. Students with accurate connected understanding first recognized that this problem involved the connections among percentile scores, the normal curve, and z-scores and proceeded from that recognition. Students without connected understanding used one of two approaches. Some tried to recall both steps as well as the z-score formula (if they were able to progress that far). Others looked for similar problems in their notes and then tried to use the identical approach in solving the test problem. Thus, students with little or no connected understanding can find ways to correctly answer these kinds of word problems. In fact, our students find problems of this type much easier than our connected understanding multiple-choice items.

Word problems can be written that require connected understanding for correct solutions. For example, students could be given all of the information found in a typical word problem *except* the question they are to answer. They could be asked to generate one or more questions that could be answered from this information. A skeleton of such a problem follows.

> From a research study, you have determined that the relationship between math scores and verbal scores on an achievement test is .60 based on a sample of 45 eighth-grade girls and .70 based on a sample of 35 eighth-grade boys. Generate three questions (in word and hypothesis forms) that could be answered using this information.

We have used problems such as this one as part of our classroom instruction but have not yet included them in formal instructional assessment or in our research.

USING CONCEPT MAPS TO ASSESS CONNECTED UNDERSTANDING

Concept maps are valuable for both summative and formative assessment of connected understanding. As instructors, we want to know how well our students understand statistics at the end of our coverage of topics and eventually at the end of our statistics course; in addition, most of us have to assign grades that reflect our judgment of their understanding. Concept maps can provide this kind of summative assessment. As importantly, though, we need to know how students' cognitive networks develop and change throughout their exposure to statistics. Concept maps can provide this formative information also. Using them formatively allows us to alter instruction to benefit our current group of students; using maps summatively benefits our current class in consolidating their learning and our future classes by indicating areas that need more instructional emphasis.

Two approaches have been used to assess connected understanding with concept maps. In the first, students draw (or create) their own maps. In the second, students complete an incomplete map. Other forms of visual representation (e.g., flowcharts) can be used with either of these approaches, although we have not explored their use in this chapter. See Jonassen et al. (1993) for an extensive presentation and discussion of the use of various kinds of maps for assessment.

Map creation

Novak and Gowin (1984) developed concept maps for use in research and evaluation in science education. Since that time, concept maps that have been drawn directly by learners have been used as measures of connected understanding. Students arrange important concepts into a map and connect them with links they label. Novak and others have developed various quantitative scoring systems for these maps. Points are awarded based on map characteristics, including number of correct propositions, levels of hierarchy, cross links, and examples.

In an introductory statistics course, for example, students could be asked to create a concept map that includes the most important concepts and their connections from a unit they have completed. Students can be given the concepts or asked to generate them. They can complete this task individually or in groups. The maps can be drawn by hand, on computers, or by arranging note cards. Other variations exist; see Ruiz-Primo and Shavelson (1996) for a summary of concept map formats and scoring variations as used in science education.

There is growing research evidence, based on expectations from theory, that the creation of concept maps can be a valid measure of connected understanding for many students (e.g., Horton, McConney, Gallo, Woods, Senn, and Hamelin, 1993; Novak and Musonda, 1991). First, research findings indicate that the maps of experts are larger, more complex, more connected (especially across neighborhoods), and qualitatively more sophisticated than those of novices. There also is evidence that the structure of students' concept maps become more like those of experts (often their instructors) with increasing educational exposure.

Map completion

Almost any map can be used as the basis for a completion, or fill-in, assessment format. The general process involves first constructing a master map. Keeping that map structure intact, some or all of the concept and/or relationship words are omitted. Students fill in these blanks either by generating the words to use (we call this format "generate-and-fill-in") or by selecting them from a list which may or may not include distractors (we call this format "select-and-fill-in"). Figures 3 and 4 are examples based on the two master concept maps found in Figures 1 and 2. Figure 3 is a select-and-fill-in map; Figure 4 is a generate-and-fill-in map. Surber (1984) may have been the first to use this approach.

We applied the select-and-fill-in approach to concept maps for our work. For a small pilot project in our graduate-level introductory applied statistics class, we created a global master map that was similar to Figure 3 but much more complex. It included 40 nodes; we removed 17.

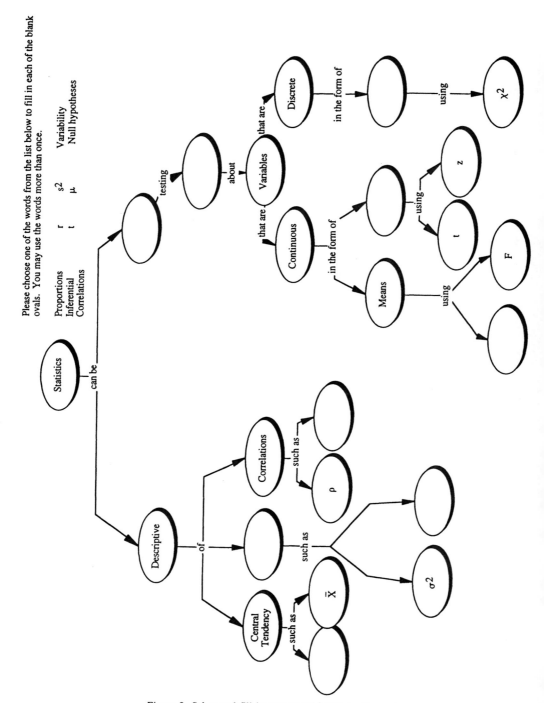

Figure 3. Select-and-fill-in assessment based on Figure 1

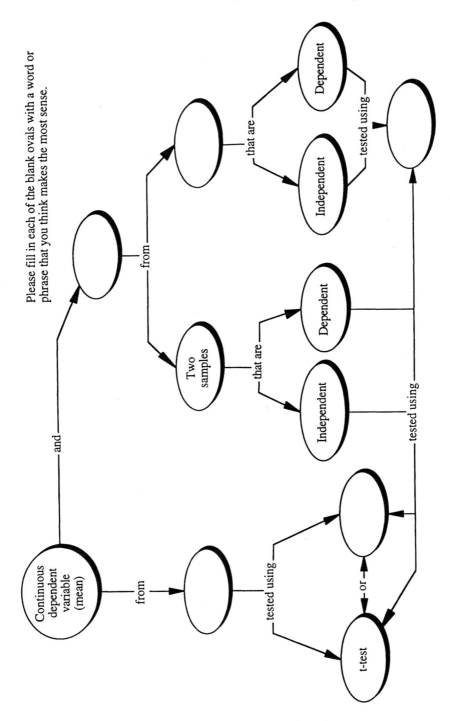

Figure 4. Generate-and-fill-in map based on Figure 2

Students were asked to complete the map at the beginning (pre) and at the end (post) of our course. They selected their answers from a list found on the map; they could reuse the words as often as needed. Map scores were not used as part of our grading scheme, and students voluntarily participated in the project.

We scored each response as correct or incorrect (although other scoring schemes are possible). We examined changes in scores from the pre to the post administration, for the class as a whole as well as for individuals.

Some of these students had taken an introductory statistics course when they were undergraduates, although in some cases that was two decades ago. Even so, at the beginning of the course, students correctly filled in, on average, 25% of the nodes. As a group, these students entered the course with some understanding (some correct, some incorrect), rather than knowing nothing.

At the end of the course, on average, students correctly filled in 73% of the nodes, a large improvement from the pretest average. In fact, most students' post scores were two or three times greater than their pre scores. We also found that these post map scores were strongly related to total points in the course; total course points were based on homework, quizzes, and tests, all of which also were designed to measure interrelated conceptual understanding.

At the end of the course, it was clear (and expected) that most students did not understand all of the important conceptual connections. As instructors, we will attend more closely to these in future classes. If we had administered these maps as a formative assessment during (rather than at the end of) the course, we could have addressed these structural misconceptions during the remainder of the course.

These patterns suggest that select-and-fill-in maps measure understanding of statistical knowledge. Students exhibited a very large gain in connected understanding of introductory statistics concepts across the course, and their fill-in scores were related to achievement in the course. Naveh-Benjamin, Lin, and McKeachie (1995) obtained similar patterns of results using a different kind of fill-in map structure.

SUMMARY AND IMPLICATIONS

From research and instructional experience, we know that many, if not most, people in the U.S. cannot engage in effective statistical reasoning, even after instruction in statistics. One viable reason is that students try to learn statistics (and mathematics) by memorizing formulas. When faced with the conventional kinds of word problems common in statistics classes (which rarely, if ever, look like problems found in everyday life or in research), they attempt to recall a formula, substitute into it, and calculate "the" correct answer. It also appears that many teachers instruct using this approach. This approach to learning and instruction posits that understanding in statistics consists of a collection of relatively isolated facts and formulas; this assumption is clearly untrue.

According to theories about schemas and mental models, as well as work on national standards and goals for mathematics, understanding requires a cognitive network of connected knowledge. These networks can be represented visually as a map of connected concepts. We use the concept map format: a map of labeled concepts connected by labeled links. This model leads easily to assessment formats based on concept maps.

There are two main approaches to the use of concept maps in assessment. The most common approach requires students to draw their own maps; the second, an approach receiving increasing attention, asks students to complete an incomplete map.

Maps can be created in a variety of ways. Students can either generate the concepts they include in their maps or use concepts that are given to them. The concepts can be put on movable pieces to encourage students to try out various structures before settling on one or more. This work can be done individually or in groups. Students generate valuable discussion about the concepts and their connections when they present and/or display their maps. We believe that this flexibility makes them most valuable as a learning tool and as a dynamic (formative) assessment tool, one that qualitatively informs the instructor about areas of understanding and misunderstanding that can then be addressed.

Asking students to generate concept maps has limitations as a summative approach to assessment. First, there is no simple and accepted scoring system for quantifying the quality of concept maps (Ruiz-Primo & Shavelson, 1996). Second, students must learn to draw concept maps. Based on our experience, this process can become tedious and frustrating since it often takes several revisions before arriving at an adequate map. Third, it takes time to generate adequate concept maps. Fourth, the process of drawing concept maps creates a high level of cognitive demand, both visual-spatial and verbal. Fifth, maps are unique to each individual since each person's cognitive understanding is unique (although, of course, there are similarities among experts' maps). Sixth, some students (and instructors) do not like to draw concept maps.

We are experimenting with map completion as both a formative and a summative assessment format. These map completion formats have several advantages. First, an easy score to obtain, and one that is understandable to others because it is familiar, is the number (or percent) of correct responses. Second, fill-in formats can be designed to be easy or difficult. Harder assessments are created by constructing more complex master maps, by omitting more words, by omitting words that are close to each other in the map, and/or by requiring students to generate their responses rather than select them. Third, fill-in formats can be administered relatively quickly and to large groups of students. Fourth, they do not require a computer for administration or scoring, although computers can be used for these purposes. Fifth, fill-in assessments may well involve a learning component since students see the map structure as they complete the task. Sixth, we have evidence that students of almost all ages, with 5 to 10 minutes of training, can and will try these fill-in structures. Seventh, fill-in maps look like they measure connected understanding; this "face" validity is an important consideration in the use of assessments in the classroom, although not necessarily a concern in research.

Another set of emerging techniques for measuring connected understanding that we have not tried to use in our statistics education work is characterized by a two-stage indirect approach. These techniques have been used primarily by researchers, often psychologists, interested in studying cognitive structures; they rarely have been used in instruction. As one example, Johnson, Goldsmith, and Teague (1995) have run a series of research studies in which students rate the degree of connection or "relatedness" between pairs of concepts. Used in statistics, for example, students could be asked to rate the relatedness between pairs of concepts selected from those found in Figures 1 and 2. Using these ratings to represent the "cognitive distance" between each pair of concepts, students' cognitive networks can be represented visually (in the form of a map) and numerically (as scores indicating the agreement between experts' and students' ratings). Both representations are easiest to obtain using computer software programs. These techniques have a number of limitations for use in the classroom. In our view, their major disadvantages are in their dependence upon technology and the problem of convincing students that the tasks are worthwhile. At this point in their development, these kinds of techniques make better research tools than classroom assessment measures.

Before selecting a format to assess connected understanding, you need to consider your model of learning and thinking and determine the purpose for your assessment. Teachers and researchers will want to use a variety of kinds of measurement for different purposes. The types of connected understanding assessment formats we have described in this chapter complement, but do not replace, other kinds of achievement assessments. Many people believe that assessment drives instruction. If so, we hope that the use of these techniques will encourage statistics teachers and researchers to consider the connections that are inherent among statistical concepts and how to encourage connected understanding in students.

Part III

Innovative Models for Classroom Assessment

Chapter 9

Assessing Statistical Thinking Using the Media

Jane M. Watson

Purpose

The goals of this chapter are (a) to address the need to assess statistical thinking as it occurs in social settings outside the classroom, (b) to suggest a hierarchy for judging outcomes, (c) to provide examples of viable assessment based on items from the media, and (d) to discuss the implications for classroom practice.

INTRODUCTION

At the end of the nineteenth century H. G. Wells claimed that "statistical thinking will one day be as necessary for efficient citizenship as the ability to read and write" (quoted in Castles, 1992, p. v). The growing emphasis, in recent years, on probability and statistics in the mathematics curriculum might be seen as acknowledging Wells's prophecy. This increased statistical content is evident in the curricular documents of many western countries: for example, in Australia the mathematics curriculum includes *chance and data*, in England and Wales, *data handling*, in the United States, *statistics and probability*, and in New Zealand, *statistics*.

Beyond merely placing technical statistical topics in the curriculum, Wells's emphasis on the needs of society in relation to statistical thinking is also reflected in curricular statements. In Australia it is found in *A National Statement on Mathematics for Australian Schools* (Australian Education Council [AEC], 1991) under headings which require students to "understand and explain social uses of chance" (p.175) and "understand the impact of statistics on daily life" (p.178). In the United States, the National Council of Teachers of Mathematics' (NCTM) *Curriculum and Evaluation Standards* (1989) exhibits the same sentiments. There it is claimed that the curriculum should provide situations so students can develop an appreciation for "statistical methods as powerful means for decision-making" (p.105) and "the pervasive use of probability in the real world" (p.109). Hence probability and statistics are suggested for inclusion in the mathematics curriculum not only because of their innate worth as intellectual topics but also because of their application in dealing with issues in wider society.

The need for statistical thinking in social decision-making is exemplified every day in the news media, where reports appear on topics as wide-ranging as politics, health, town planning, environmental control, unemployment, sport, science, and attendance at cultural events. If evidence of the need for statistical literacy is found in the media, then the media is also an ideal vehicle to provide initial motivation for the study of statistics, applications of specific topics in the curriculum during instruction, and items for assessment in the final stages of learning. It is the contention of this chapter that the validity of context-based instruction, in this case using the media, is only ensured if the context is employed at all stages of the learning cycle.

The issues related to the assessment of statistical thinking are not only the province of mathematics teachers. As other curriculum areas acknowledge the importance of the analysis of data and risk in relation to their subject matter, the need for statistical literacy is seen in many differing contexts. Recent curricular reform in Australia illustrates this trend. *A Statement on Studies of Society and Environment for Australian Schools* (AEC, 1994a) includes a process strand, "Investigation, Communication and Participation," which states the following expectations for students:

> Students gradually build up the skills involved in research, processing data and in interpreting or applying findings.... This is a foundation for predicting possible solutions to the problem, constructing hypotheses, considering other approaches to inquiry, and designing suitable methods for gathering and organising information. Sources of information are assessed for their authenticity and credibility (p.11).

These expectations rely very heavily on skills of statistical thinking, and associated specific outcomes are listed in the accompanying *Profile* document (AEC, 1994b). Similar statements are found in curriculum documents in science, health and physical education, and technology. Teachers of many subjects in the school curriculum are facing the task of teaching and assessing statistical thinking in some context.

TARGET SKILLS AND LEVELS OF ACHIEVEMENT

Once we accept that statistical thinking in social contexts is an important part of statistical education, it is necessary to describe the associated skills and their levels of complexity. This will assist teachers in structuring learning experiences and planning the related assessment. The skills required to interpret stochastic information presented in society, often in the form of media reports, can be represented in a three-tiered hierarchy: (a) a basic understanding of probabilistic and statistical terminology, (b) an understanding of probabilistic and statistical language and concepts when they are embedded in the context of wider social discussion, and (c) a questioning attitude which can apply more sophisticated concepts to contradict claims made without proper statistical foundation. These skills represent increasingly sophisticated thinking and are consistent with models of learning from developmental psychology (see for example, Biggs & Collis, 1982, 1991; Case, 1985; Watson, Collis, Callingham & Moritz, in press). Each will be considered in turn.

Tier 1: Basic understanding of terminology

At the first stage of the hierarchy there are the skills related directly to specific topics in the curriculum; these are generally taught in a conventional fashion with students creating and analyzing their own data sets. At various levels of the curriculum the topics include percentage, median, mean, specific probabilities, odds, graphing, measures of spread, and exploratory data analysis. All can be taught without reference to social issues, and detailed discussions of their teaching and assessment are found elsewhere in this volume. If, however, students ever express the query, "Why do we have to learn this?", the evidence from the media is very easy to find. In fact, asking students to find it for themselves may be the most motivating avenue to pursue. If used at this level media extracts should be very straightforward. Searches are likely to produce examples such as the following short story ("Weight guidelines," 1995), which mentions percentage and average.

> United States federal guidelines on healthy weights for men and women are too lenient and may be encouraging Americans to weigh too much according to a new study. The study found that women of average weight in the US had a 50 per cent higher chance of heart attack than did women weighing 15 per cent below average.

Although used initially to motivate basic definitions, discussion can lead to the next level of the skills hierarchy.

Tier 2: Embedding of language and concepts in a wider context

Once students with some rudimentary statistical concepts in hand are exposed to the media, a second need—to read and interpret written reports, rather than just perform computations—becomes important. Some students who have excelled in the traditional symbolic aspects of the mathematics curriculum resist the requirement for reading, interpreting, and writing when mathematics is presented in non-symbolic contexts. In all aspects of mathematical thinking, however, the need for application, interpretation, and communication skills is being recognized (e.g., Schoenfeld, 1992). The specific necessity to tie statistical and literacy skills together is acknowledged in curriculum documents around the world; for example, in Australia the language skills for young children are prescribed in two outcomes for the early and late elementary years (AEC, 1991): "use with clarity, everyday language associated with chance events" (p. 166), and "make statements about how likely are everyday experiences which involve some elements of chance and understand the terms 'chance' and 'probability' in common usage" (p. 170). These objectives demonstrate the need to begin early in tying probability and statistics to everyday experience.

As students mature, the need for language interpretation skills is no less important and the media readily supply examples. Newspaper headlines could provide a motivation for the need to relate likelihood to everyday events, a basis for class discussion of relative likelihood, or an assessment task. Students could be asked to rank the following five headlines from least likely to most likely, with justification: (a) "Willis looks sure bet for Treasurer," (b) "Freak proves it's no fluke," (c) "Blacks' shocking jail odds," (d) "Stingrays grab second chance," and (e) "Ghost of a chance something funny will happen."

The understanding of risk is important in many decision-making situations and other curriculum areas which demand an appreciation of risk analysis. This involves making assumptions of student understanding based on work in probability in the mathematics class. Consider, for example, the following extract (Ewing, 1994) that might be used in a mathematics class or a science class.

> There is a one-in-10,000 chance that an asteroid or comet, more than two kilometres in diameter, will collide with Earth in the next century, killing a large proportion of the population, according to space scientists.... An article in the British journal *Nature* says the

risk is great enough to justify a space surveillance system that would warn scientists of approaching objects and allow them to deflect them with nuclear explosions.... A person living in the United States has a greater chance of dying from a comet impact (one in 20,000) than being killed in a flood (one in 30,000) or a one-in-100,000 risk of death from a venomous bite or sting.

Again the need is seen for an understanding of probabilistic language if sense it to be made of the article.

Statistical language offers the same challenge when placed in context. The extract on weight noted earlier also can be used at this level to reinforce understanding. At this second level of sophistication, more than just basic definitions are needed to be successful. It is necessary to recognize these in other contexts and be able to make sense of claims which are made. Since on most occasions statistics are presented correctly, the requirement at this level is to understand and interpret statistics in order to draw conclusions and make decisions.

Tier 3: Questioning of claims

At the highest level of the statistical thinking hierarchy, students possess the confidence to challenge what they read in the media. It sometimes happens that claims are made without proper statistical foundation, either inadvertently or purposefully. Whether there is an intention to mislead or just insufficient information, students need to be made aware of the expectation that they must constantly question conclusions. The specific questioning skills required at this level are exemplified by Gal (1994) and relate to sampling, the distribution of raw data, appropriate use of statistics, graphs, causal claims made, and probabilistic statements.

The astronomical example given earlier provides more than just an opportunity to interpret risk in terms of probability. It gives the opportunity to develop (or assess) student skills in interpreting stochastic information and questioning the motives of people who use it. The purpose of the article was to convince politicians or finance-granting authorities to fund a $67 million-plus project to list all potentially threatening asteroids large enough to precipitate a global catastrophe. Questions of how the claimed probabilities were obtained are not answered in the article and the statistically literate reader would be highly suspicious of the figures, given the desire of the scientists to fund their pet project. Thus, to move students from a situation where they automatically believe everything they read in the media to one where they intelligently question data and claims is an important aspect of statistical literacy.

Similar skills are important in relation to an understanding of samples and populations and their use in media reports. Small samples, samples without a mention of their size, and non-representative samples should all be treated with suspicion and results based on them considered with scepticism. An extract like the following ("Decriminalise," 1992) can be used to elicit higher order analysis skills from students in this area.

Some 96 percent of callers to youth radio station Triple J have said marijuana use should be decriminalised in Australia. The phone-in listener poll, which closed yesterday, showed 9,924—out of the 10,000-plus callers—favoured decriminalisation, the station said. Only 389 believed possession of the drug should remain a criminal offence. Many callers stressed they did not smoke marijuana but still believed in decriminalising its use, a Triple J statement said.

Although the sample size is very large, the question of the population represented by the listening audience and the voluntary nature of the phone-in procedure lead to scepticism about the claims that could be made on the basis of the report.

For classroom purposes it is important to appreciate the increasingly complex nature of the thinking involved as students move from developing a basic understanding of statistical terminology and concepts in a mathematical context, to understanding and applying them in a wider social context, to questioning their use by those who may wish to mislead members of society. Assessment of these skills goes hand-in-hand with their teaching and learning. There is no reason why the media cannot be used as a basis for assessment as it may have been earlier for initial motivation or classroom discussion in conjunction with learning concepts.

BASIC ISSUES IN ASSESSMENT OF STATISTICAL THINKING

Although there can be many purposes for assessing statistical thinking, including research or the development of state- or nation-wide norms, the emphasis here is on assessment to inform teachers for instructional purposes and students for progress reports. It is hence the formative or summative nature of the assessment which is of interest. Even with these narrower aims in mind, however, the scope for using innovative methods of assessment is wide (see e.g., Webb, 1993).

For classroom teachers the major result of any type of assessment should be the evaluation of whether unit objectives have been met and the informing of future instruction, be it for remedial work following a test or for planning next year's teaching sequence. In this context the types of media-based items introduced in the next section can provide formative evaluation of the teaching program. Such items may be administered in testing situations or used for group discussion. In either situation it is the level of response achieved which should be the concern of the teacher. Can students interpret statistical information in context and can they question suspicious claims? If not, then it will be necessary to adapt teaching practice to assist students in achieving higher levels.

A secondary necessity for teachers may be to provide a summative measure of students' performances. For this purpose schemes such as those suggested for problem solving in the NCTM *Standards* (1989) or by Charles, Lester and O'Daffer (1987) may be modified for use. This involves assigning increasing integer values as students achieve higher levels of sophistication in their responses. This approach will be outlined briefly in relation to the items presented in the next section.

It will be seen that the items introduced here are of the *open-ended* type that require students to provide written (or oral) responses rather than to choose from a selection of teacher-composed alternatives. This method is used with the intent of allowing students to demonstrate statistical understanding and questioning ability which would not be possible in a multiple-choice format. This is consistent with Schoenfeld's (1992) comments in terms of the assessment of mathematical thinking. In supporting the use of items of the type suggested here he says,

> To state the case bluntly, current assessment measures (especially the standardized multiple-choice tests favored by many administrators for "accountability") deal with only a minuscule portion of the skills and perspectives encompassed by the phrase *mathematical power*... (p. 365).

At the third tier of statistical thinking assessed here, indeed a high degree of mathematical power is being demonstrated.

It will be noticed also that computation is not required as part of the assessment used here; this is generally the case when media reports are used for assessment. There are two reasons for this. First, the media seldom provide raw data upon which computations can be performed. Second, the interpretative skills being assessed generally involve the understanding of concepts rather than computation. The purpose of assessing interpretive skills is to discover if students can move to higher levels of cognitive functioning than are generally required to perform computations. The skills here encompass the communication of ideas as well as the recognition of meaning in context (NCTM, 1991).

There is little doubt that it is easier to make decisions about calculations than about many of the responses shown in this chapter. The requirement to judge levels of sophistication in statistical thinking means that teachers must understand the concepts well enough not to be fooled by students who may be grasping at straws to fill in a blank space on a piece of paper. There has been some concern by mathematics educators (e.g., McGregor, 1993) that once writing about mathematics becomes part of assessment in the subject, teachers are likely to reward quantity rather than quality. Students who write many lines, very neatly, are apt to be given more marks than those who are succinct and perhaps not as tidy. Therefore, a big responsibility falls on the teacher to develop interpretive skills to a high level as well.

FORMATIVE ASSESSMENT WITH MEDIA ITEMS

In this section two examples will be presented to indicate how articles from newspapers can be used to assess students' interpretive statistical thinking. The items were used initially as part of a large survey of student understanding of statistics presented in social contexts which was carried out with 670 students in Grades 6 and 9 in Tasmania. The responses of some of these students will be used as a basis for discussion of the items presented here. The responses will illustrate the variety and level of understanding shown by students in these grades. The students who were administered the Media Survey had had mixed classroom experiences with probability and statistics. Some of the Grade 6 students had been exposed to lessons on experimental probability, and some Grade 9 students had had a unit on statistics. The variety of responses provide a range from what could be described as naïve common sense to more sophisticated interpretive skills. In some cases intuition is used; it may or may not lead to acceptable outcomes.

The examples represent two content areas from the statistics curriculum, each presented in a different context, and offer the opportunity for students to question a contention made in each report and reach the third tier of the statistical thinking hierarchy. The content areas represented are graphical representation, using a pie chart in an economic context, and sampling, in a social science context. The use of media extracts related to probability is considered elsewhere (Watson, 1993).

There is no reason why these items cannot be used in teaching as well as assessment. The emphasis here, however, is on the type of achievement which can be expected when the media extracts are used for assessment. Webb (1992) suggests four components of the assessment process which are helpful in following through with the procedure: situation, response, analysis and interpretation. These will be used in the following examples to consider the formative assessment possible in relation to statistical thinking. *Situation* will be interpreted here to mean the context set by the question asked. The physical situation associated with the administration of the item may vary from classroom discussion, to group work, to individual written responses and will not be discussed for each example. The *response* phase will be illustrated with responses from the Tasmanian survey. The *analysis* will be based on levels associated with the latter two tiers of the statistical thinking hierarchy introduced earlier. Finally, *interpretation* will connect the analysis back to classroom practice.

The framing of questions seeking to elicit understanding of media extracts is not an easy task. The objective is to allow for various interpretations but perhaps to guide at some stage of the questioning protocol. Various forms of questions were trialed in the pilot work leading to the items used in this chapter. Those presented here may not be considered perfect, but they did produce a wide range of responses. Teachers should be aware of the necessity to think carefully about the questions they ask in order not to bias the responses.

Graphical representation

Representation of data graphically is one aspect of statistics which has been in the mathematics curriculum for a long time and it might be expected that students are familiar with various types of representation. The work of Pereira-Mendoza and Mellor (1991) and Curcio (1987) reveals, however, that students are likely to have beliefs about the features of graphs that are different from what is expected. It is important to reinforce the understanding of graphing in the high school years; and as the following item shows, there is scope for testing the highest level of statistical thinking: the questioning of information presented.

Situation

The headline and pie chart shown in Figure 1 appeared, with an accompanying article, on page 1 of the *Australian Financial Review* (Webb, O'Meara & Brown, 1993). The questions in Figure 1 ask the meaning of the pie chart and whether there is anything unusual about it. It is assumed that the Tasmanian responses are similar to those that would be obtained in most developed countries.

It is obvious to those who understand the principle that a pie chart represents 100% of the quantity under consideration that there is something wrong with the pie chart in the figure. It is of great interest whether students will notice this mistake.

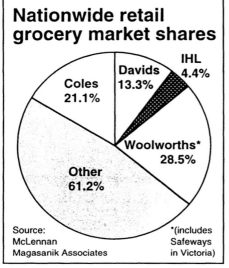

Figure 1. Pie chart question

Response

Of Grade 6 students, there are quite a few who say they do not understand the pie chart and leave the question blank. This type of response diminishes greatly by Grade 9. The following responses would likely be acceptable to most teachers for explaining the meaning of the pie chart: (a) "It tells us what % they have sold," (b) "That Coles got 21.1%, Davids 13.3%, Woolworths 28.5%, other 61.2%," and (c) "It shows who out of the 5 markets who [sic] has the most share of the grocery market shares." These responses have achieved the second tier of the statistical thinking hierarchy.

When open-ended responses are allowed, however, many are found to be lacking in some respect. The response, "It says how well those shops are selling," has interpreted the situation correctly but has it contained enough information about the percentage shares of the market? The response, "It shows Nation wide retail grocery market shares," represents a tautology, merely repeating the information in the title of the pie chart. The response, "It means how many people go there," equates market share with number of shoppers. Is this a reasonable assumption? It also alludes to quantity ("how many") rather than percentage. "I think it is about all the people who have taken out shares" misunderstands the term *share* in the context and illustrates the need to appreciate the language of the application before being able to make adequate statistical interpretations. Whereas the response, "How much money is being made," has interpreted the context correctly, it has not mentioned the relative information for the different groups in the pie chart and again emphasizes quantity rather than percentage. Is this student just being careless or is the understanding not present?

In a classroom it may be possible to answer these questions directly—a big advantage over a large-scale impersonal survey. If the answers were part of a whole-class discussion of the pie chart's meaning, they would probably be greeted with an encouraging, "Yes, that's right. Now what else can we tell from the chart?" How the answers should be handled in a written format is more difficult.

Moving to the second part of the question in Figure 1, the recognition of the error in the pie chart represents a higher level of functioning: the third tier of the statistical thinking hierarchy. Those who can answer the second part correctly can give concise functional responses to the first part. Not all students who can answer the first part adequately, however, can produce the higher level response needed for the second part.

When they are asked if there is anything unusual in the pie chart many students give "No" responses, and many others fail to answer the question at all. Instead of recognizing the percentage error in the pie chart, students' lower-level responses suggest instead that there is something "unusual" within the chart itself. The first five responses represent aspects not related to the statistical task in hand: (a) "It's cut into all different shapes," (b) "It doesn't say what 'Other' shops are," (c) "They are all decimal like 21.2, 13.3, 28.5, 61.2," (d) "The black part," and (e) "I can't figure out why Woolworths have a star and the rest don't." The next three responses mention aspects which are related to the statistical representation but these are not the significant unusual feature to one who understands fully the creation of a pie chart: (a) "Other is bigger than the rest," (b) "Coles is one of the smallest market shares," and (c) "The heading doesn't fit in." The "other is bigger" response is the most common answer of this type. At the level at which this part of the question is directed, these responses would be considered not to have engaged the question; that is, they do not achieve a level of functioning higher than that required for the first part of the question.

The recognition of the percentage error in the pie chart can be noted in two ways: (a) "The percentages add up to 128.5. They should equal 100!!" and (b) "Where it has Other, it says 61.2% and the percentage of that section on the pie is less than 50%." It could be argued that response (a) is more sophisticated in its realization of the incorrect total. Response (b) may not be accompanied by this realization but be the result of perceived inaccuracy in drawing the sections of the pie.

Analysis

Whereas on first glance, the question about the meaning of the pie chart may appear to be a relatively straight-forward application of this form of graphical representation in a particular context, the responses above show that partial understanding is possible from several perspectives. Hence to achieve a response which fully satisfies the first two tiers of the statistical thinking hierarchy—to show a basic understanding of the statistical representation and appreciate how that representation is embedded in a wider context—requires a fairly high level ability to relate ideas together. Intermediate level responses are exhibited above and it is possible to suggest an ordering as follows.

- No engagement with the item:

    "Don't know."
    "I was absent when we learned pie graphs."

- Single facet of the item accessed in response:

    "That others is all the other grocery market."
    "I think it is about all the people who have taken out shares."

- Appreciation of some more complex aspects of the pie chart representation:

    "It says how well those shops are selling."
    "It shows the percentage of people in the population who have taken out shares..."

- A full understanding which relates pie chart information in context:

    "It tells us what % they have sold."
    "That Coles got 21.1%, Davids 13.3%, ..."
    "It shows who out of the 5 markets has the most share of the grocery market shares."

Students responding at this last level could be claimed to have satisfied the second tier of the hierarchy of applying statistical understanding in a social context. Reaching the third tier is reserved for those who could go further and question the percentage figures given in the pie chart. There is therefore a final level of achievement for this item.

- An ability to correctly question claims made:

    "The percentages add up to 128.5. They should equal 100!!"
    "Where it has Other, it says 61.2% and the percentage of that section on the pie is less than 50%."

The importance of ranking intermediate responses is to recognize the need to raise levels of response if given classroom opportunities and to provide a basis for giving summative feedback if required. To merely classify responses as *wrong* or *right* may mean that chances are missed to encourage progression to higher levels.

Interpretation

It is important for teachers to appreciate the degree of complexity involved in the intermediate responses which lead up to what would be considered a fully adequate reply. It is the acknowledgment of this complexity and the rewarding of movement from one level to the next that is part of the contribution a teacher can make to the improvement of performance by students. A class discussion which leads students from single facet replies to more complex responses representative of the second and third tier will assist students to form the connections to be able to produce higher level responses.

There are teachers who would claim that the graphical representation created by a pie chart is not fully understood until the second part of the item in Figure 1 is answered correctly. In that case the importance of including in the assessment an error which embodies the essence of the pie chart concept is immediately seen. Those who answer the first part adequately have shown an appreciation of some of the features of a pie chart. The necessity to represent a whole with 100% was not a feature mentioned by many in answer to the first question, but it is encouraging that many did recognize the lack of the property when asked to explore further. The second part of the pie chart item illustrates the importance of the third skill in the hierarchy: a questioning attitude which can apply the more sophisticated concepts to contradict claims made without proper statistical foundation.

Sampling

The importance of sampling as part of the statistics curriculum has been recognized more recently than graphing; it is now specifically mentioned, however, in current curriculum documents (e.g., AEC, 1991; NCTM, 1989). Sampling activities (e.g., Friel & Corwin, 1990; Landwehr, Swift & Watkins, 1987) are now suggested for both primary and secondary students, and the importance of using an appropriate sampling procedure is stressed in relation to instances of misleading claims being made (e.g., Watson, 1992). Deficiencies in students' basic understanding of sampling are also being documented (e.g., Rubin, Bruce & Tenney, 1991), which further supports the need for challenging assessment items in the area.

Situation

The extract shown in Figure 2 appeared in the "That's Life" (1993) column of the *Hobart Mercury*. The first part of the question associated with the report (see Figure 3) asks students for any criticisms they may have of the claims in the article. It is meant to be open-ended to allow for other than statistical criticisms if students focus on them. The second part provides a more specific opportunity to consider the handgun claim for the whole of the United States based on a sample from Chicago.

The concept to be assessed is the relationship between a sample and a population. As neither of these terms is mentioned in the article, students must have both the basic understanding of the terms and the ability to recognize them in a social context when other words, such as *poll,* are used. The essence of the questions, however, is to gauge if students have reached the third tier of statistical thinking where they can question the claim about an inappropriate population based on the sample. Because of this desire to assess the highest level of statistical thinking, it was necessary to word the questions in a way to avoid the words *sample* and *population*.

ABOUT six in ten United States high school students say they could get a handgun if they wanted one, a third of them within an hour, a survey shows. The poll of 2508 junior and senior high school students in Chicago also found 15 per cent had actually carried a handgun within the past 30 days, with 4 per cent taking one to school.

Figure 2. Newspaper article about sampling

Would you make any criticisms of the claims in this article?

If you were a high school teacher, would this report make you refuse a job offer somewhere else in the United States, say Colorado or Arizona? Why or why not?

Figure 3. Questions for students

Response

Many students in both Grades 6 and 9 say "No" to both questions. Of the Grade 6 students in the Media Survey who express criticisms in answer to the first part of the question, none focus on the sample-population question. This may not be surprising since most Grade 6 students would not have been exposed to very sophisticated work on sampling. It is interesting, however, to note that the criticisms are mainly aimed at the implications of the claims, and not the content of the claims: (a) "Rules in USA schools are terrible!" (b) "They shouldn't have hand guns at all," and (c) "I think that the owners of the gun shops should be fined for letting the kids buy them." Few Grade 6 students in any way question the report. The criticism, "I doubt that they could get one within an hour...", is probably related to the student's social experience, which would be very different from Chicago. The student who responds, "Because they could be lieing [lying]," is questioning the veracity of the evidence given by the Chicago students, which is indeed an issue when conducting surveys.

Many Grade 9 students make comments similar to Grade 6, but there are some who do appreciate the sample-population problem: (a) "Yes that with only 2508 students interviewed they could have only found [them] in [the] rough part of town," (b) "They only interviewed people in Chicago," (c) "The articles poll is done in a city with crime not in a country where the % could be 1," and (d) "They are saying 6 in 10 United States high school students. But actually it is 6 in 10 Chicago high school students." The first two responses focus on the sample without reference to the population but in each case it appears that the relationship is implicitly understood. The second two responses explicitly state the difficulty in the article.

The second question in Figure 3, related to teaching in another place than Chicago, offers a further chance for students to pick up the error in the article by suggesting a scenario which explicitly states a different geographical location for comparison. This added probe results in some Grade 6 students noticing that the sample from Chicago may not be representative of Colorado or Arizona, even though they had originally ignored the reference to the United States as a whole: (a) "No because Colorado and Arizona might not have children with guns," (b) "No because it only talks about Chicago," and (c) "I'd have to find out more about it, this story might not be true for all high schools." Similar responses are given by a larger proportion of the Grade 9 students, again by many who had missed spotting the difficulty in their answers to the first part of the question. Some of these responses are stated in a more sophisticated manner: (a) "No because just because Chicago is like that doesn't mean other places would be," and (b) "No because the whole of the United States wasn't surveyed, so we don't know that the handgun situation is the same throughout the United States."

It is also of interest to consider the responses made by students who continue to miss the main point of the questions. There are some who probably do not take the question seriously: (a) "I don't want to be a teacher," and (b) "No because I'd have a gun too." Many accept and agree with the claims being valid outside the sample area. Some of these responses may be influenced by the students' experiences of the United States on television: (a) "Yes because I wouldn't want to take the risk of getting shot," and (b) "Yes because Americans are insane." Some also say they would teach in other areas but for reasons not associated with the sampling problem: (a) "No, because only 4% are taking them to school," and (b) "No because it doesn't say that these kids fired the gun, they were just showing off. Being tough."

Analysis

In terms of defining levels of response appropriate for this question, it is the second part of the question which provides the foundation for building a picture of student understanding. Many students can provide a statistically acceptable answer to the second part but not the first. The reverse however does not occur; those who can answer the first part can always answer the second. Categories as given below seem appropriate for this item.

- No engagement with concepts:

    "I'd have a gun too."
    "Rules in USA schools are terrible."

- One peripheral statistical point:

    "They could be lying."
    "Only 4% are taking guns to school."

- Recognition with geographical cue (second part of question):

    "No because it only talks of Chicago."
    "No because the whole of the United States wasn't surveyed ..."

- Recognition without geographical cue (first part of question):

    "They are say 6 in 10 United States high school students when..."

Interpretation

It should be noted that this item is structured differently from most items used for assessment. Usually the first part of a question is easier, leading on to a more sophisticated second part, as in the previous pie chart item. The reason for structuring the sampling question in the way it was presented mirrors the manner in which the article might be used as a basis for classroom discussion. A teacher beginning a session on sampling might present the article and ask students to discuss the first question. It is likely that many responses, similar to those given here which were off the point, would come up during discussion. The teacher would then focus the class on one of the high level responses if it arose or move on to the second question as a means of leading students to an appreciation of the sampling problem. It is unlikely a teacher would ask the first question and if no adequate answers were forthcoming then ask, "What is the relationship of the Chicago poll to the claim about United States high school students?" Such a question would not allow for individuals to discover the relationship for themselves. The second question in this item might allow that to happen.

It is important that assessment items are designed to reflect the manner in which teaching and learning takes place. Hence it should not be considered impossible to ask a more difficult question first in some circumstances. This is especially so when the highest level of skills is being assessed.

SUMMATIVE ASSESSMENT WITH MEDIA ITEMS

If summative assessment is to be carried out in conjunction with the use of the items introduced above, then a marking scheme based on the levels in the "Analysis" section for the items can be developed. The increasing degree of sophistication can be associated with increasing integer values. This is a holistic scheme, which is a variation on that suggested by Charles et al. (1987). For the first part of the pie chart question assigning *0, 2, 4, 6*, to the four categories of response with the possibility of odd numbers for answers, displaying slightly less understanding or quality would be appropriate. If one were to score the entire item including both questions in a hierarchical fashion, it would appear reasonable to give the complete answer (including the "128.5" response for the second part an *8* and including the "61.5" response) a *7*. For the sampling item it would appear reasonable to assign quantitative scores of *0, 2, 4,* and *6* , respectively, for responses in this hierarchy.

The assigning of marks described here may seem relatively simplistic, but much effort has gone into the definition of levels reflecting increasing achievement. It is unlikely that summative assessment for students would consist only of numbers but also would include comments explaining why the numbers were assigned. This is the method employed by Garfield (1993) in reporting on project work in statistics.

IMPLICATIONS

It has been the purpose of this chapter to examine the levels of thinking required to interpret statistical information and claims made in social settings (and reported in the media), to relate these to the assessment of interpretive statistical thinking, to provide examples of how an assessment scheme operates, and to consider some of the classroom issues associated with such assessment. The goal is to achieve third-tier statistical literacy for all students in terms similar to those stated by Wallman (1993) in an address to the American Statistical Association.

"Statistical Literacy" is the ability to understand and critically evaluate statistical results that permeate our daily lives—coupled with the ability to appreciate the contributions that statistical thinking can make in public and private, professional and personal decisions. (p.1)

There are several implications for the classroom teacher which arise from the need to assess interpretive statistical thinking and from the examples used in this chapter. Webb (1992, p. 667) claims that assessment which is integral to instruction embodies four features. Each of these is paraphrased in terms of the context of this chapter. First, the teacher must understand the structure of the statistical thinking hierarchy and use this structure to define expectations for learning. Second, the teacher must be sensitive to the processes students use to learn critical statistical thinking, the stages of development, and the processes available to facilitate this contextual statistical learning. Third, assessment is a process of gathering information about a student's knowledge about statistical thinking, about the structure and organization of the tiers of that knowledge, and about a student's cognitive processes, then giving meaning to the information obtained. This is the goal of the "Analysis" and "Interpretation" sections for the items introduced in this chapter, and it would be expected that teachers would acquire the ability to structure other assessment in a similar fashion using that model. Fourth, assessment employing items such as described in this chapter is used to make informed decisions about methods of instruction. Such instruction should be based on current information available about what a student knows and about what a student is striving to know. The aim is to assist each student in achieving higher levels.

Several other issues warrant mention with regard to the assessment of statistical thinking in the classroom. One relates to the number of students who are assessed in relation to a single outcome. Many curriculum documents (e.g., AEC, 1991; NCTM, 1989) suggest cooperative group work and report-writing in connection with objectives such as those discussed in this chapter. The assessment of such work, both within classrooms and on a larger scale, has not been addressed widely in the mathematics curriculum, let alone with respect to statistics. Assessment questions relate to the assignment of sub-tasks within a group, the degree of input provided by each participant, the quality of the final product produced by the group, and individual learning which has taken place as a result of the group activity. All of these questions will need to be addressed for items which begin with media extracts and involve cooperative work leading to assessment. Because much of the work in society which leads to media reports is the result of the efforts of teams of researchers, pollsters, and so forth, it is relevant to develop assessment techniques valid for teams as well as individuals in this context.

The combination of motivated students, well-informed teachers, relevant content and a useful scheme for assessment should ensure that the higher order thinking required for statistical literacy is achieved for most if not all students by the end of secondary school.

Chapter 10

Assessing Students' Statistical Problem-Solving Behaviors in a Small-Group Setting

Frances R. Curcio and Alice F. Artzt

Purpose

The ability to interpret and predict from data presented in graphical form is a higher-order skill that is a necessity in our highly technological society. Recent recommendations from the mathematics and science education communities have therefore stressed the importance of engaging learners in real-life statistical tasks given in a setting that will promote effective problem solving. Since the small-group setting has been shown to be a fertile environment in which problem solving can occur, we have used that setting for engaging students in data analysis tasks. However, there is a dearth of ideas related to how to assess students' behavior, thinking, and performance in such a setting. The purpose of this chapter is to describe a framework for assessing students' problem-solving behaviors on a graph task as they work within a small-group setting.

INTRODUCTION

Statistical problems that require data analysis are not encountered only when one studies statistics, but rather, permeate other disciplines as well. For example, mathematics, the physical sciences, the biological sciences, and the social sciences all employ data analysis as a tool for solving problems. Yet, although many current assessment items require that students calculate specific statistics (e.g., measures of central tendency, measures of dispersion) or locate specific information (e.g., reading a graph or table), opportunities to present statistics as a problem-solving process have been ignored (Garfield, 1993; Friel et al., this volume).

The small-group setting has been demonstrated as an effective environment for enhancing mathematical problem solving (e.g., Artzt & Armour-Thomas, 1992). Such a setting provides a safe environment in which students can brainstorm, share, and discuss their interpretations related to higher-level cognitive tasks. Small-group interactions can contribute to developing mathematical power in statistical reasoning (Curcio & Artzt, in review). Only recently have educators begun to identify assessment techniques in such a setting (e.g., Hibbard, 1992; Kroll et al., 1992; Miller, 1992), recognizing the importance of assessing students' contributions to a group task (MSEB, 1993).

This chapter describes our efforts to design and apply a framework for assessing statistical problem-solving behaviors of students as they work in a small-group setting. This framework is demonstrated in the context of assessing the problem-solving of a group of middle school students working on a graphical task. It should be noted that the chapter deals with two separate but related facets of assessment that emerge in such a context: assessment of the *process* of problem-solving of a group of students, and assessment of the *quality* of the solution of a *graphical* problem. We also discuss integrative assessment techniques.

The chapter is organized in four parts. First, issues in assessment of performance on graphical tasks are noted, and the specific task given to our students is described. Next, a framework for assessing problem solving in a small group setting is described and adapted to the graphical task presented to the students. The use of this assessment framework is then demonstrated by applying it to the actual work of a group of four students, and the types of information about students' work processes that we were able to obtain are illustrated. Lastly, implications for instruction and assessment are discussed.

GRAPHICAL TASKS AND THEIR ASSESSMENT

Interpreting and analyzing data are problem-solving processes that are essential for dealing with information presented in many different forms, including but not limited to graphs and tables. Visual displays of data are found in reports and in the media, often being used to make decisions or to determine whether to support or reject arguments. Recognizing trends, extracting patterns, and extrapolating from data are higher-order problem-solving components of data interpretation and analysis.

There are several frameworks for analyzing graphic interpretations (Guthrie, Weber, & Kimmerly, 1993; McKnight, 1990; Pinker, 1990). This chapter uses a basic framework for analyzing students' responses that was presented in Curcio (1981, 1987). This framework involves notions of *reading the data* (i.e., extracting information explicitly and directly from the graph), *reading between the data* (i.e., combining and comparing data), and *reading beyond the data* (i.e., extrapolating and predicting from the data).

One of the most frequently used visual displays is the time series plot (Tufte, 1983). Found in science and social studies documents, the time series plot is a line graph of data recorded over a period of time. Using this type of display enables scientists and social scientists to examine patterns, recognize trends, extrapolate from the data, and make predictions. Recognizing the importance of such a visual display, a higher-order, open-ended, free-response task employing such a graph was designed to be the focus of a small-group assessment. The structure of the task is similar to extended constructed-response item-types found on a recent National Assessment of Educational Progress (NAEP; Dossey et al., 1993). The task was designed to be of interest to and mathematically worthwhile (MSEB, 1993) for middle school students, and can serve as a prototype for other graph tasks.

The task

Students were given a table and a time series plot containing information about the average time of sunset from June to December (see Figure 1). The task was presented in two parts. The first part required that the students, working alone, read the two displays of the data, think about the data individually for a few minutes, and write their interpretations of the given information.

Part 1: Read the following two displays of information. Working alone, write as many things as you can about the information given.

AVERAGE TIME OF SUNSET

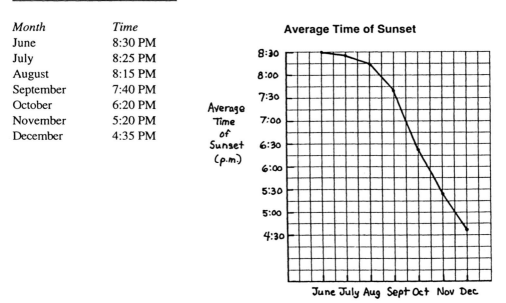

Month	Time
June	8:30 PM
July	8:25 PM
August	8:15 PM
September	7:40 PM
October	6:20 PM
November	5:20 PM
December	4:35 PM

Share your interpretation of the information with the members of your group. The recorder of the group is to write a statement that represents the group's interpretation.

Part 2: Work with the members in your group. Using the line graph entitled, "Average Time of Sunset," draw a picture of what the graph would look like if it were to continue from January to May.

Figure 1. The two parts of the graphing task

Successfully completing the task depends on students' awareness of changes in the amount of daylight hours during the different months of the year and relies on their everyday experiences for a reasonable solution.

After completing their individual written interpretations, students were organized into small groups of four to share their ideas and record their agreed upon interpretations. One student in the group was the recorder and kept a written account of the ideas.

For the second part of the task, as the students worked in the same small groups of four, they were asked to extend the graph by drawing a picture of what the graph would look like if it were to continue from January to May. This part of the task required students to recognize and discuss trends in the data, relate their experiential knowledge about the amount of daylight hours during the different months of the year to the task at hand, and integrate and translate this information into a time series plot for the remaining months of the year, January through May. The students' problem-solving behaviors on the task were recorded using a framework adapted from Artzt and Armour-Thomas (1992). A full description follows.

A FRAMEWORK FOR ANALYZING STUDENTS' PROBLEM-SOLVING BEHAVIORS

Polya's (1945) conception of mathematical problem solving as a four-phase heuristic process (i.e., understanding, planning, carrying out the plan, and looking back) has served as a tool for investigating problem-solving competence. More recently, these categories have been expanded and examined by researchers who have begun to identify the importance of metacognition in mathematical problem solving (e.g., Garofalo & Lester, 1985; Schoenfeld, 1987; Silver, 1987) and in assessing higher-order thinking in mathematics (Baker, 1990). These researchers identified monitoring and self-regulation as metacognitive behaviors that are crucial for successful problem solving in mathematics. Small problem-solving groups provide natural settings for discussion in which interpersonal monitoring and regulating of members' goal-directed behaviors occur. These may well be the factors that are responsible for the positive effects observed in small-group mathematical problem solving and so these are the factors that one would wish to encourage and assess as students participate in small mathematical problem-solving groups.

The results of a study that examined problem-solving behaviors within the context of small groups (Artzt & Armour-Thomas, 1992) suggest that the continuous interplay of cognitive and metacognitive behaviors that occurs among members of successful problem-solving groups mirrors the thoughts and behaviors of expert problem solvers working alone. Students working in small groups have the opportunity to communicate about mathematics as they discuss and develop problem-solving strategies.

Artzt and Armour-Thomas (1992) developed a framework for analyzing the interactions among the members of small groups. They argued that, as students work within a small group to solve a mathematics problem, their behaviors range from *talking about the problem* (a metacognitive behavior), to *doing the problem* (a cognitive behavior), to *watching or listening* to other students talking about or doing the problem. At times, some students may be *off task* and exhibit none of these behaviors.

In the remainder of this section we analyze in detail the many elements that may be part of these four general categories of such group problem-solving behaviors. This analysis is useful as it outlines the range and diversity of behaviors, skills, and interactions that may have to be the subject of assessment efforts by teachers who want to promote effective group problem-solving by their students.

Below we explain how these four categories of group problem-solving behavior pertain to graph tasks in general and to the specific graphical task presented earlier. The framework below relates to observable behaviors but also remarks about possible underlying mental activities (mainly cognitive and metacognitive) and related problem-solving heuristics that presumably occur during the work process. In reading through our analysis, the readers are asked to assume that they are observing a group of students and have to record any observations on a form such as the one described in Figure 2 (and discussed in detail later on).

We should note that the order of categories below is not necessarily the sequence in which the problem-solving behaviors may occur. Several approaches or steps that could have been used to complete the task are also mentioned, in order to raise issues that go beyond the particulars of the task used in this chapter.

Talking about the problem

This category of behaviors is comprised of six possible types. Three are listed in this section: understanding, analyzing, planning. Three additional behaviors are listed later as comprising parts of "doing the problem": exploring, implementing, verifying; these may also be part of "talking about the problem," and can be considered as either cognitive or metacognitive, depending on the types of actions taken or statements made.

Understanding. This subcategory is indicated when students make comments that reflect attempts to clarify the meaning of a problem. In general, when students have to interpret or create a graph, they must understand what the axes represent and the relationship between variables (e.g., what are the independent and dependent variables). They must also be familiar with the meaning of and preferred uses of specific graphical representations (e.g., bar graph, stem-and-leaf plot, circle graph).

In our graphing task, the students must understand that a graph is to be drawn that extends through the remaining months of the year. They must understand that they have to do three things in order to complete the task:

1. A rectangular grid must be drawn.
2. The grid must have an x-axis containing the months, January, February, March, April, and May.
3. The y-axis must contain the average times of sunset, which they are to figure out.

Analyzing. This subcategory relates to moments when students make statements revealing that they are trying to simplify, reformulate, or analyze a problem. Reading between and beyond the data are reflected in such statements. In a general graphing task, students may recognize patterns or trends in the data. They may show that they are familiar with the context of the data and use their background knowledge or experiences to help make sense of what they read in the graph or what they will represent in the graph.

In our graphing task, the students may analyze the problem in the following ways:

1. Looking at the patterns from the given data they can notice a trend in the change of average time of sunset from one month to the next.
2. They can think about their personal experiences with sunset hours in the months from January through May.
3. They can think about what the average time of sunset might be if the pattern were to continue and compare that with their experiences.

Planning. This subcategory refers to cases when students make statements about how to proceed in the problem-solving process. In a task that requires an interpretation of a graph, they may decide to look for patterns or trends in the data. To create a graph, they may decide which variables belong to the appropriate axes. They may approximate appropriate ranges for the data and intervals to use. They may plan what type of graph they will use.

In our graphing task, if the students attempt to plan an approach for solving the problem, they may do the following (note that the plans might not be ones that will lead to a successful solution):

1. Draw a coordinate grid with the months from January through May across the x-axis.
2. List the average times of sunset on the y-axis. (The range of these times differs.)
3. Find the difference between each successive average time of sunset and look for a pattern in the differences.
4. Continue the pattern of differences to calculate the average time of sunset in the remaining months by subsequent subtractions.
5. Use the pattern of differences to create a pattern of subsequent additions to find the remaining average times of sunset.
6. Use personal experiences of all group members to get the remaining average times of sunset.

Doing the problem (or Talking about the problem)

This category is comprised of four types of behavior. The first, "reading," is predominantly cognitive (though of course involves metacognitive processes). The three remaining types, as mentioned above, may belong either in this category ("Doing") or in the previous one ("Talking"), depending on the types of actions taken or statements made. They can involve either cognitive or metacognitive processes.

Reading the problem. At some points students may read the problem or data or listen to someone else read. The metacognitive aspect of this reading behavior can only be recorded when students verbalize their understanding of what they have read.

Exploring the problem. Exploration as a cognitive activity alone often results in disorderly, aimless, and unchecked wanderings. When exploration is guided by the monitoring of either oneself or one of the group members, that behavior can be categorized as exploration with metacognition. Such monitoring leads to self- or group regulation of the exploration process, thereby keeping the exploration controlled and focused. For example, when students are called upon to interpret a graph they may mistakenly interchange the variables and come to an incorrect conclusion about the trend they notice. If they or their group members do not take the time to consider whether their interpretation makes sense, they are exploring without metacognition. The same metacognitive behaviors are necessary when students embark on the creation of a graph without prior planning. Without monitoring, this exploratory form of graph creation can lead to improper use of axes, inappropriate intervals and range for variables, and the use of inappropriate graph types.

The graphing task lends itself to some guessing and testing because the differences in the data on successive months of average times of sunset do not form a consistent pattern. While the average time of sunset from June through December does steadily decrease, it does not do so in a discernible pattern. This means that students usually begin estimating what the continuation of the pattern could be. Without using monitoring, this exploration might lead to an incorrect continuation of the decrease in average time of sunset. Another unmonitored exploration might lead a student just to continue the line formed by the given graph. When monitored, these types of explorations could lead to students realizing that a change in direction must be made because their experiences do not support their results.

Implementing. At some points students may devise a plan for solving the problem, and may attempt to implement the plan. If the students perform this implementation systematically with monitoring and regulating (metacognitive), they are likely to discover that either the plan was good and has led to a reasonable solution, or it was faulty and needs some adjustment. If the implementation is unmonitored (i.e., cognitive level alone), however, students may follow through on the implementation leading to an incorrect and unreasonable solution.

Verifying. An effective verification requires students to examine their final response or solution and check that the answer makes sense (metacognitive). When interpreting a graph, students must compare what they think the graph is saying either to their knowledge of the individual pieces of data or to their life experience and knowledge about the data and its source. After having created a graph, they must check to see that the representation conforms to what they know to be true about the data.

In our graphing task, the students must determine whether the average times of sunset they have calculated are consistent with their personal observations of average times of sunset in the months of January through May. If they have tried to continue the pattern in the given data as it is, or in the reverse direction, they may try to verify (metacognitive) their results by checking their successive subtractions or additions (cognitive).

Watching and listening. These behaviors cannot be categorized as either cognitive or metacognitive because they are not audible. However, students must be willing and able to listen and watch each other in order for an exchange of ideas to take place.

Off-task behaviors. These behaviors cannot be categorized as either cognitive or metacognitive because they are not related to the problem-solving task.

Based on the framework described above, we have developed an instrument (see Figure 2) to aid us in our quest, as researchers, to understand the problem-solving behaviors of students as they work in small groups (Artzt & Armour-Thomas, 1992). Yet, teachers who have familiarity with problem-solving heuristics may readily use the instrument. A teacher could systematically examine the behavior(s) of each student in a group being observed for about one minute and note the proper category or categories of behavior. Important anecdotal information that further describes the students' behaviors could be added at the bottom of the chart. For example, the teacher might wish to point out which students were responsible for making what Schoenfeld (1985) refers to as "executive decisions." These are the key decisions that can account for the success or failure of the group work, such as insightful plans for how to solve a problem, or the acute observation that the problem solution is going in the wrong direction.

	Talking About the Problem						Doing the Problem				Watching and Listening	Off-Task
Name of Student	UNDERSTANDING	ANALYZING	PLANNING	EXPLORING	IMPLEMENTING	VERIFYING	READING	EXPLORING	IMPLEMENTING	VERIFYING		
Student A												
Student B												
Student C												
Student D												
Anecdotal Information												

Figure 2. The Assessment Instrument

ASSESSING THE STATISTICAL PROBLEM-SOLVING BEHAVIORS OF STUDENTS

In order to assess problem-solving behaviors properly, a variety of evidence about student performance must be considered (NCTM, 1995). To get an enhanced picture of students' understandings, one can first assess students' individual work and then examine their within-group behaviors in light of that work. Very often, there may be a discrepancy between a student's level of performance when working alone and his/her level of performance when working within a group. When this occurs, the teacher must consider the effect the group is having on the student and make changes when necessary.

In our project, to determine children's levels of knowledge prior to beginning the group task, their written statements for the individual part of the graphing task (i.e., interpreting data) were examined in terms of the categories of *reading the data, reading between the data,* or *reading beyond the data.* As an example, below are responses from four fifth graders, Chuck, Dennis, Garin, and Razzie, about what they thought was communicated in the time series plot:

Garin: The graph shows the average time of sunset per month.
Dennis: October sunsets one hour later than November.
Razzie: In December the sun sets earlier.
Chuck: In the summer it gets darker later than the winter about a difference of 4 hours.

Garin's statement simply restates the title of the graph; it reflects *reading the data.* Dennis and Razzie use comparatives in their statements (i.e., "later than" and "earlier," respectively), reflecting *reading-between-the-data* comprehension. Chuck introduces "summer" and "winter" ideas into the graph; his statement reflects *reading-beyond-the-data* comprehension.

It is interesting to note that when the students were given the chance to share their ideas and agree on a group interpretation, they chose Chuck's statement, which reflected the highest level of comprehension. Specifically, Razzie, who volunteered to be the recorder wrote: "Everybody in this group agrees that the sun sets an estimate of four hours later in the summer than winter. "

For the second part of the graphing task (i.e., creating a sketch of a graph), the group was given one sheet of unlined paper with the following instructions:

Using the line graph about "Average Time of Sunset," draw a picture of what the graph would look like if it were to continue.

The children worked for approximately 15 minutes to complete the task. The first seven minutes of their problem-solving behaviors recorded by an observer can be found in Figure 3. To give the reader a frame for interpreting this chart, an overall picture of the group's problem-solving efforts is presented below. Examples of the specific events that contributed to the checks entered for each student are then described.

From Figure 3 it can be seen that during the first seven minutes of their group problem-solving session none of the children was off task; each of them at some time watched or listened to one another. However, although it seems apparent that each of the students was involved in the problem-solving process, the quality and quantity of their contributions varied. For example, Garin took the lead in solving the problem. He engaged in almost each type of problem-solving behavior at both cognitive levels. That is, he was instrumental in coming up with ideas about how to do the problem and he participated in actually doing the problem.

Razzie, on the other hand, seemed to remain on the outskirts of the group solution. She made several efforts to try to understand what the problem was asking for, and aside from one instance of analyzing the problem, she remained in an exploratory cognitive phase. Chuck and Dennis were integrally involved in the solution of the problem. They were attentive to the ideas of all, as is evidenced by the multitude of instances in which they were watching and listening, and they both took part in analyzing the problem and implementing the ideas about how to approach the problem.

Name of Student	Talking About the Problem						Doing the Problem				Watching and Listening	Off-Task
	UNDERSTANDING	ANALYZING	PLANNING	EXPLORING	IMPLEMENTING	VERIFYING	READING	EXPLORING	IMPLEMENTING	VERIFYING		
Chuck	√	√√		√	√√√√		√		√		√√√√√√	
Dennis		√√√		√	√√√√	√	√		√√√	√	√√√√√√√	
Garin	√√	√√	√	√√	√√√	√	√		√√√	√	√√√√	
Razzie	√√√	√		√			√	√√			√√√√√	

Anecdotal Information

Chuck, Dennis, and Garin were responsible for doing most of the problem solving while Razzie, for the most part, watched and listened. Garin was the only one who came up with several plans. Chuck, Dennis, and Garin all worked on implementing the plans. Chuck, through careful monitoring of the implementation, made the important observation that, if they were to continue following the pattern they would arrive at an unrealistic solution. This observation caused Garin to revise the plan and set the problem solution on track.

Figure 3. Recordings of Problem-Solving Behaviors During Small Group Work on Graphing Task

When examining some of the specific comments made by the students, one can get a greater understanding and appreciation for the contributions made by the students and the coding decisions made by the teacher.

Examples of reading and understanding the problem:

In the beginning of the group session, after reading the problem, several students could be heard trying to understand the problem. Razzie asked, "What do they mean to continue for all the months?" Garin asked, "No, but listen, how can you continue, it ends in December?"

Examples of analyzing the problem

The students then began to *analyze* the problem and arrive at some ideas. Garin said, "Maybe you have to start from January all the way through May." Dennis added, "There has to be some kind of pattern there so we could figure it out." Razzie noticed, "Check this out. Look, 5 minutes after in July the sun sets 5 minutes before it did in, um, June."

Examples of exploring the problem

To get better acquainted with the problem, the group got involved in a bit of *exploration*. Chuck said, "So wait, estimate the times here (pointing to the displays in center of the desk). 8:30, 8:00, wait a minute...It goes by half hours so just figure out like..." Razzie tried to complete Chuck's thought, "Chuck, 8:15, half an hour, a whole half an hour."

Returning to understanding and analyzing

Then they returned to trying to *understand* how to go about doing the problem. Chuck said, "Don't do it by this [the table], do it by the graph, it said the graph." Garin questioned, "We have to draw pictures, right?" After examining the data on the graph for a short while, Garin, Chuck, and Dennis engaged in further *analysis* of the problem. Garin stated, "Wait, see January starts a new year. See every month it goes down about an estimate of 15-20 minutes. OK, an estimate of 20 minutes. So if you get 8:30, if you do it 8:30 and you keep on subtracting 20 all the way up till January, you'll get your estimate." Chuck added, "The average time goes down about half an hour." Garin responded, "No it doesn't, look because..." Dennis interrupted, "No, if we're going down a half hour, this is 15 minutes and then you just have to keep going down." During this time Razzie was still trying to *understand* the problem. She asked, "Can I ask you a question? Are any of you sure what you're talking about?...Very sure?"

Examples of planning and further analysis of the problem

Finally, Garin announced a *plan* (although faulty), "So let's see, if you start in, in June, it's 8:30 and you keep subtracting maybe 20, keep on subtracting 20 you will probably get your answer for every month, maybe." But Dennis and Chuck were not about to accept the plan without further *analysis*. Dennis noticed, "But look, this is 5 and the next one is 10. But then the next one is like no pattern." Chuck added a point of information, "In May it's like 9 o'clock." Dennis added, "There has to be some kind of pattern, but, between here is 5 and between here is 10."

Examples of implementing the plan

Despite the lack of evidence of a consistent pattern, Chuck, Dennis, and Garin began to *implement* Garin's plan for repeated subtractions. Luckily, Chuck was monitoring the implementation and made the important observation, "By the time we get to [inaudible] we're going to be down to like 3 o'clock. It gets dark at 3 o'clock? Be for real." Based on this statement, the boys began to make adjustments to their previous *implementation*. Dennis said, "No, because then you have to like start making the time go on higher." Garin responded, "Oh, that's right. It goes like this. Listen, 8:30, subtract, [pause] then it will probably go down." Dennis replied, "The difference in between the time gets larger and it also gets smaller. It starts decreasing." And Chuck adds again, "In May it gets dark at 9 o'clock."

Examples of making a new plan and final implementation

The above discussion sparked Garin to come up with a new *plan*. He suggested, "Look at this. Would you listen? Look it goes down every 5, 10, 35. So why don't you just do this--subtract there from 5 and do it like, this is like your middle number. You got it?" The group used the remaining time to implement this improved plan. As Razzie began to draw the grid, impatiently, Garin grabbed it from her to construct the graph. Chuck and Dennis closely monitored his work and Razzie looked on.

With detailed data from the individual and group efforts (of Chuck, Dennis, Garin, and Razzie, in our case), the teacher has access to a wealth of information to provide for thoughtful assessment. These kinds of data should be further evaluated to yield a more complete picture of students' level of statistical problem-solving. This integrative evaluation is discussed in the next section.

INTEGRATIVE ASSESSMENT

It is interesting to note that the results of both the individually agreed-upon interpretation of the sunset data and the completed group graph (see Figure 4) reflect the highest level of comprehension. That is, the statement that the students agreed on for a description of the data was Chuck's statement, "In the summer it gets darker later than the winter about a difference of 4 hours, reflecting reading beyond the data.

Overall, students' interaction appears to have brought the group to a level of reading beyond the data, illustrated in their final product in Figure 4. Contrasting the levels of comprehension of the students' individual statements with the problem-solving behaviors they exhibited when they worked within their small groups, however, reveals some consistencies as well as some inconsistencies.

In his individual work, Chuck was the only student who made a statement reflecting the highest level of cognitive performance--reading beyond the data. Similarly, within the group, he also performed at a high level. Not only was he actively engaged in all aspects of the problem-solving process, but he was the one who made the keen observation that following Garin's plan would lead to a senseless solution. This is an example of the type of monitoring of a problem solution that is characteristic of expert problem solvers.

Dennis, when working on his own, made a comparative statement reflecting the ability to read between the data. His active involvement and understanding of the problem solution within the group showed a similar level of comprehension as that which he demonstrated while working alone.

Razzie made an attempt to make a comparative statement–reading between the data–when working on her own. Yet, she really did not seem to grasp the full meaning of the problem during her group work. She made several attempts at trying to understand what the problem was asking for, but as evidenced by what seemed to be her aimless explorations, she never attained more than a very low level of comprehension.

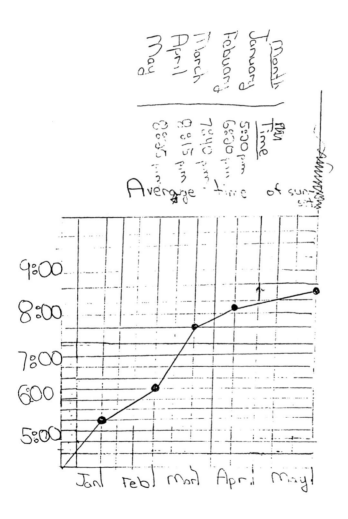

Figure 4. Solution to Graph Task

In his individual work, Garin was the only student who made a statement reflecting low level cognitive performance--reading the data. In extreme contrast, within the group he performed at a higher level than all of the students. That is, he had the greatest number and variety of metacognitive behaviors, and he was the only one who suggested any plans for how to approach the problem (see Figure 3).

The teacher and the students have much to learn from these results. For the teacher, instructional decisions may be made for individuals as well as for the group. For the students, reflecting on their problem-solving behaviors can reveal a great deal about their own strengths and weaknesses as individuals and as members of a group.

IMPLICATIONS

Two key purposes of assessment are to provide students with an accurate window into their emerging capabilities, and provide teachers with a broad picture about students that can help them make informed instructional decisions. The approach to assessment of a group problem-solving process described in this chapter can contribute information that can serve both the teacher and the students. Although only two sources of evidence were available in the assessment method described here (i.e., results of the individual and group parts of the graph task), additional sources of information are available to teachers (see other chapters in this volume) and should be incorporated into a holistic assessment of students' knowledge and performance.

Making Instructional Decisions

The approach to assessment illustrated in this chapter can and should serve as an integral part of instruction–helping teachers to make instructional decisions about content, grouping of students, events in one group that should be discussed by the whole class, and so forth. When teachers notice, for example, that metacognitive problem-solving behaviors such as regulating, monitoring, and checking occur within a group, they need to bring these behaviors to the attention of the class. An example is Chuck's important statement (see "Examples of implementing the plan" above) that caused his group to redirect its efforts.

Likewise, opportunities for children to recognize the importance of questioning the reasonableness of their problem-solving processes or solutions should be provided. For instance, Garin served as a rich source of ideas and suggested a strategy for implementing a plan, but this would have led to an unreasonable solution without Chuck's observation. Garin and the other students should be helped to realize that deficiencies in monitoring and checking may impede their problem-solving performance.

Teachers and students should be made aware that inconsistencies in students' individual and within group behaviors may occur, and need to be alert to the need to make adjustments in group formation. For example, a more suitable group assignment for Razzie should perhaps be sought as the group setting Razzie encountered in the group described here did not appear to contribute to her understanding of the task. Decisions about changing group assignments should be made, for example, based on the group's social and academic performance (Artzt, 1994).

The graph task presented to the fifth graders was both engaging and challenging for them. Based on their successful completion of the graphing task and their level of interest, the teacher may find it worthwhile to extend this experience. For example, a follow up activity may require that the students collect their own data about daily AM and PM temperature, average monthly precipitation, or accuracy of weather forecasts. At this point, students may be ready to design their own display (e.g., table, line graph), describe why they selected it and explain what information they expect to be able to extract from it. They could use graphing software to experiment with different types of displays, and compare the usefulness of the information each conveys.

Student self-assessment

Another way to use the results of the graphing task is to empower students by engaging them in the process of self-assessments (Paulu, 1994). Getting students to look at their own problem-solving abilities and behaviors when working alone or in groups supports their development as autonomous learners. To maximize students' awareness of their own problem-solving behaviors, the assessment method described here could be adapted for use by the students themselves. Information students record on it can later inform a group discussion focused on events or processes (in both the cognitive and metacognitive domains) that contribute to the final product. Such a discussion could explore the contribution of ideas suggested or actions taken by specific members of a group, as well as highlight the contribution of processes such as monitoring or questioning the reasonableness of solutions.

Through a self-assessment process, students can become more aware of their contributions to the group in comparison to their individual levels of comprehension. For example, it seems that Razzie may have had more to offer than what was exhibited in the group. It would have been useful for her to determine what was it about this particular group setting that caused her to become withdrawn. Garin, on the other hand, could benefit from reflecting on his somewhat impulsive approach to devising and implementing plans. He should be made aware of the importance of regulating and monitoring his behaviors. These issues, of course, are general and their resolution can contribute to students' general problem-solving skills.

In summary, it is important to reiterate that the time series plot used in this exploratory work is representative of many other graphs and plots with which students should become familiar (Curcio, 1989; Friel et al., in this volume; Macdonald-Ross, 1977). Graphical tasks in general lend themselves to higher-level problem-solving behaviors, as they challenge students to integrate, apply, and transform what they already know. The many substeps described earlier as being part of the process of solving a statistical problem in a group context are quite general, and many also pertain to individual problem solving. We have attempted to demonstrate how individual assessment techniques can be supplemented by techniques that monitor students' problem-solving behaviors as they work within small groups, to yield rich information that can inform instructional decisions. As educators now realize, using multiple sources of evidence about students' strengths and weaknesses provides a more complete picture of their capabilities (NCTM, 1995).

It is not yet clear how interacting in small groups helps improve individual statistical problem-solving ability. Yet, it is clear that by using an assessment instrument that places value on higher-level problem-solving behaviors, students are more likely to be sensitized to the importance of higher-level cognitive and metacognitive processes, and can be enlightened about techniques for improving their own individual graph comprehension. Assessment techniques that focus on group processes and behaviors can potentially provide students and teachers alike with a wealth of rich information about ways to improve statistical reasoning skills.

Acknowledgments

This chapter is based on a paper entitled, "The Effects of Small Group Interactions on Graph Comprehension of Fifth Graders," presented at the Seventh International Congress on Mathematical Education, Quebec, 1992. Thanks to Professor Eleanor Armour-Thomas who assisted in the design of the study and the analysis of the data reported herein.

Chapter 11
Assessing Student Projects

Susan Starkings

Purpose

The main purpose of this chapter is to offer practical advice to teachers who want to use projects in their courses. In this chapter some examples of projects are given, two assessment models are explained, and teachers' experiences are described. The project models and examples described have been used with students of 14 to 18 years of age, but can be adapted for younger or older students as well

INTRODUCTION

Assessment plays a major role in education, possibly more so now than ever before. A key role of assessment is the diagnostic process—by establishing what students have learned, it is possible to plan what students need to learn in the future. Project work is a method of allowing students to use what they have learned in statistics classes in a practical context. It is this practical application of projects that make them such a useful part of the learning process.

Although project work may look easy, a brief introduction with this way of working will show how demanding it really is for both teachers and students. Students must make connections between one piece of learning with another. They have to transfer the skills acquired in statistics to other areas such as science and geography, and vice-versa. They have to familiarise themselves with a wide range of information. This is much more demanding than learning one isolated fact after another. Integrated work of this kind is often the best preparation for higher education and future employment.

Many forms of assessment, such as assignments and learning activities, are written by teachers for students, i.e., teachers do the thinking and students act on their instructions. Educational psychologists have shown that learning is not like walking in a straight line or climbing a ladder. We do not all learn one thing, then another, then another, in the same order. It is more accurate to compare a student's learning to doing a 3D jig-saw puzzle or making a complicated model rather than climbing a ladder. We put in a couple of pieces, then we may do another section altogether. Sometimes we make a connection we had not seen previously, and occasionally we link two sections together that had not been apparent before. Project work allows students to connect various pieces of knowledge together that suits a solution to a chosen problem.

This chapter examines issues that arise in the process of designing, doing, and reporting a project. These issues relate both to the students doing the project as well as to the teacher (who serves as both facilitator and assessor). The chapter is divided into seven sections:

1. Starting a project and initial advice
2. Staged assessment
3. Teaching structure during project completion
4. Presentations of projects
5. Teachers' experiences
6. Feedback
7. Implications

STARTING A PROJECT: INITIAL ADVICE

The project

Any project involves a student, or group of students, working under the guidance and direction of a teacher. A project involves planning and developing a schedule of study and outcomes to be achieved over a period of time usually longer than that of an individual assignment. Implicit in the project are educational ideas of deep learning as opposed to surface learning. The students are not required to only produce a single outcome, but are required to link multiple, and often distantly related, prior learning tasks. The role of the teacher throughout the project is one of a facilitator of each student's learning strategy.

It is important that both teachers and students know precisely what is involved in doing project work. The students should be familiar with the assessment structure that will be used by the teacher to assess and grade their project. The next section gives some assessment models that could be used for this process. The students should be given a copy of the model marking scheme (or grading guidelines or scoring rubric) that is chosen to formally evaluate their project.

Individual or group projects

Project work can be carried out individually or in groups. The formation of groups can be carried out in one of two ways: either students form their own group or the teacher identifies the group membership. Both of these methods have their own merits and drawbacks. Students who form their own groups tend to choose friends with similar backgrounds and objectives. This may detract from the rich learning experience to be gained by groups formed by students with different approaches to learning. Kelly (1978) advocates that for students the "furthering [of] their social and personal education" is an essential feature of group project work.

Teachers can form groups in a variety of ways. It could be done randomly by allocating a number to each student and selecting these numbers from a hat or by using random number tables to select groups. If this method is adapted some useful statistics can be discussed and elucidated during the selection process. One of the most useful methods used is to compose groups by individual student ability. In order to use this technique this method obviously assumes that the teacher knows his/her students. Teachers who previously used this method recommend that the group contain a mixed ability range; otherwise one may well find that a high flier group or a very weak group often materialises. (Weak groups present problems when—or if—verbal assessment is given.) Another advantage of a mixed ability group is that the most able students can be used as mentors to the least able. It is well known in educational circles that we learn by doing and by teaching others. Hence the most able students can serve as mentors to enhance their own learning as well as helping their fellow students.

Project choice

Careful choice of a project is required; otherwise teachers may find that the project does not proceed very well and in some cases not succeed at all. Whether students choose projects on their own or in conjunction with their teachers, clearly defined objectives for the projects are essential.

Teachers vary in their opinion of projects and in particular the assessment of these projects. A great deal depends on whether projects are introduced for examination purposes or for instructional reasons. If the projects are part of an examination syllabus the assessment structure should be followed as stated in the syllabus. When projects are not part of a formal examination, an appropriate model for assessment can be used. These projects can be assessed in stages or at the end, depending on the teacher's preference.

Teachers have found that a successful project begins with a clear problem scenario, with precise goals identified. From this a detailed outline of work can be produced. For example, a group of 16-year-old students decide to investigate their fellow students' smoking habits. This in itself is a project beginning, but further classification is needed. Some projects that have been successfully completed by students in the United Kingdom (UK) are:

- Predicting the Top 20 Pop Records in the Music Charts
- The number of cashiers required at a supermarket
- Students' examination marks
- School dinner meal selection
- Students' smoking habits
- Newspaper readership
- Students' perception of gambling
- Simulation using random numbers
- Comparison of photograph sizes in different newspapers.

This list is by no means exhaustive, and many more projects can, and indeed should, evolve from students' own interests. Students should be encouraged to collect and assess data arising from an area of interest and to explore different ways of organising and representing it. It is not essential for students to collect their own data; successful projects can be completed by using data from other sources. Two examples of this are as follows:

Example 1:

> A charity fun run took place on 11th August 1994 and the following data was obtained on each of the runners: Name, gender, age, time taken to complete the race, amount of money raised, and charity to which money is to be paid. Use the data stored in the "computer file xxx" to produce a report for the local press and for the Charity Fun Run Newsletter.

Statistics on the amount of money raised for each charity, the number of male and female runners, and other characteristics could be produced; it was up to the students to decide how to present the information in terms of graphical, numerical and analytical form.

Example 2:

The file "student-data" contains 99 fields of information, collected via questionnaire, on students who have just enrolled in a university computer studies course. Use this data to answer a question of your choosing about these students. The responses to the questionnaire could be examined in many ways, but you are asked to consider one aspect of the data in depth. Some areas of interest are as follows:

1. What is the pattern of academic qualifications obtained and is it related to age and gender?
2. Can an overall score for mathematics experience be constructed? If so, is it related to age or gender?
3. What is the "catchment area" for the course? That is, which geographical regions do new recruits come from? Are there any special characteristics of recruits from certain regions?

These are just some examples that were given to the students. They were also given a copy of the questionnaire and an explanation of all the fields concerned in the data file.

Example 1 is of a prescribed nature and would allow students to decide what statistics to use and to produce their own personal report, whereas example 2 encourages the students to decide on an area of interest to look at and decide what data from the file is to be extracted. These two examples must analyse previously collected data. If this is not practical or feasible in your organisation the best way is to encourage students to collect their own data. Both of the above examples assume students have access to the data electronically and can make use of relevant computer software packages for statistics. Simple data sets can also be arranged for manual computations.

STAGED ASSESSMENT

Assessment can take many forms; for example, a project can be assessed at completion or may be assessed in stages. Different models of assessment may be used. How these are assessed vary as to how the models are structured and the relevant weightings/grades that are assigned to the various parts of the project. A good motivation factor is to assess the projects in stages since this gives students an indication on their progress and induces them to continue with the work.

The purpose of staged assessment (or assessment in stages) is to provide feedback for students at various points throughout the project. This enables students to attain the maximum benefit and guidance throughout the project period. Consider the following problem:

A ferry boat sells a variety of different tickets, such as one-day return, cheap (reduced) day return, weekly ticket, monthly ticket and one-way tickets. The ferry company wants statistics collected for a management report.

The first part of the project is to identify what statistics the company requires and how these are to be collected. A grade could then be given for this first analysis section of the problem. If the student has left out some of the analysis (for example, the times when cheap day tickets operate), the teacher can provide this information and the student can proceed to the next stage.

The student then has the opportunity to attain the maximum mark possible for the next stage and so on. The next stage may be to design a "data capture" form to collect these statistics. Since the student has a detailed scenario to work on s/he can proceed to design the form with all the information present. Students can then improve their projects as part of the ongoing process. This is an example of a formative approach to provide helpful and detailed feedback about a student's progress. Caution should be applied when giving feedback, especially for staged assessment, that the giving of inflated predicted grades in order to sustain optimism and motivation by some teachers will raise as many problems as it solves.

"Motivated experience results from the exposure of learners to tasks which enables them to satisfy particular needs and hence gives rise to feelings of satisfaction when the objective is attained." (Walklin, 1991).

Staged assessment models

The two models we describe below, i.e., the ADIE model (Analysis, Design, Implementation and Evaluation) and the 4P model (Project log, Project report, Practical development and Presentation), have marks allocated for the four stages. The ADIE model was designed to be used by individual students, whereas the 4P model was originally designed for group projects.

The 4P model has a 15% weighting attached to the presentation, whereas the ADIE model does not have a specific grade allocated to a presentation. If the ADIE model is to be used, an additional mark of 10 could be allocated to a presentation, using the 4P Model's criteria.

A teacher must decide what type of assessment structure is to be used. What follows is a brief introduction to 2 models that have been developed and could be used. It may be more appropriate to use some of the above ideas and adapt them to suit the project work undertaken and facets to be covered. Note that both models allow for assessment of the quality of four stages of work.

The ADIE model

Each problem needs to be analysed first in order to design what data needs to be collected. A good project starts by analysing the problem and addressing the issues of interest or concern. Design of data collection and techniques to be used in data collection follows the analysis. The statistical routines and methods are then carried out with an evaluation of the whole process at the end. Hence, each section could be graded separately, giving feedback to students in stages. If required, the project could be graded at completion using this model. The ADIE model was developed by the Northern Examinations Board (1988) and can be used with the weightings given below. These allow for a maximal score of 40, but could be proportionally changed to be on a 0-100 scale.

Analysis (max. 10 marks)

Full, or near complete, logical breakdown.	9, 10
Reasonably full, clear and well thought out breakdown.	7, 8
Fairly clear thought out attempt.	5, 6
Partially successful attempt.	3, 4
Limited attempt.	1, 2
Not attempted.	0

Design (max. 8 marks)

Accurate, detailed. Follows closely from analysis. Revised as, and if, necessary in light of other stages.	7, 8
Reasonably accurate, detailed. Analysis considered. Some revision undertaken if needed.	5, 6
Adequate attempt showing some links with analysis.	3, 4
Some listing of resources.	2
Limited attempt.	1
Not attempted.	0

Implementation (max. 14 marks)

Follows closely the design, uses appropriate techniques with skill and understanding to produce a good solution.	12-14
Mostly follows the design and appropriate techniques used with reasonable skill and understanding.	8-11
Some linking to design and techniques used with some understanding.	5-7
Techniques applied with some success.	3,4
Some techniques undertaken.	1,2
Not attempted.	0

Evaluation (max. 8 marks)

Clearly relates solution to the problem. Shows a good understanding and appreciation of the solution.	7,8
Reasonably clear reference between the solution and the problem and some appreciation of the solution.	5,6
Some linking of solution to problem.	3,4
Mainly concerned with technical aspects.	2
Statement of what has been done.	1
Not attempted.	0

Each stage above can be assessed and graded separately. The solution can, and often does, involve several statistical techniques being employed. For example, a study taken on consumer food expenditure in a local shop may involve the collection of the amount spent by each customer and what items were purchased. The solution may be to produce a time chart of daily takings at the shop, the average amount spent per day by customers, and a bar chart of the number of customers per day. These are by no means the only statistics that could be undertaken but serve as an example to demonstrate what a solution could contain.

The 4P model

The 4P model was developed at South Bank University (Elliott & Starkings, 1994) and is assessed in each of the 4 stages. Suggested weightings are: Project Log 20%, Project report 25%, Practical development 40% and Presentation 15%. The project log is assessed on an individual basis, thus allowing for individual members within groups to be assessed this way. The assessment will take into consideration the individual student's involvement in the group and his/her individual effort and contribution to the overall project. The mark is to be made on continuous assessment, determined by the supervising teacher during the project's time span. In making a judgement the supervisor should refer to the content and accuracy of the student's individual logbook, which each group member keeps during the project life time. When grading the student's project log and the group writing project, the following should be taken into account:

Project Log:
a. The individual student's effort and commitment.
b. The quality of the work produced by the individual student.
c. The student's integration and co-operation with the rest of the group.
d. The completeness of the logbook.

Written Project Report:
a. Introduction.
b. Project specifications.
c. Statistical techniques used and calculations.
d. Solutions to the problem.
e. Recommendations and conclusions.

In addition to these components, practical development (computational steps and other "technical" activities) should be marked with regard to the written report. The presentation of the project to teachers and fellow student groups of the same institution should also be marked. Points to be considered in these areas are:

Practical Development:
a. The group's investigation of the practical aspect (as evidenced in the report).
b. Integration of the practical development with the rest of the project.
c. The group's/individual analysis and design of the problem.
d. The group's/individual attempt at practical development.

Project Presentation:
a. General quality of the presentation.
b. Integration and teamwork.

c. Interest, content and originality.
d. Use and quality of the statistics.

This model allows for staged assessment, but obviously parts are linked and may provide a more suitable model for an end of project grading. These models are an indication of what can be done to assess projects and provide a framework on which teachers can develop their own project work.

PRESENTATIONS OF PROJECTS

The presentation of projects is not always a criteria that is applied in assessment, particularly one of a statistical nature. A structure that the author has developed in the UK highlights certain areas that can be used to assess the verbal presentations of the student's projects. One of the main purposes of the presentations is to help students develop and improve their communications skills. Students need to be able to communicate technical or numerical data in everyday life. Students need to be able to explain and describe statistical methods and results, and this is a skill that is not very often tested in syllabuses (course plans). The structure shown below could be used to assess the non-statistical aspects of a presentation. The presentation structure considers and awards grades on an equal weighting as follows (each element gets an equal weighting):

a. Relationship to the audience: Appropriate material used, clear message and recommendations that are meaningful.
b. Use of supporting materials and useful aids to communication: Use of appropriate media for demonstration, handouts, etc.
c. Structure: Introduction, middle, summary, and conclusion.
d. Handling questions: Thoughtful and honest responses, appropriateness of answers and convincing arguments.
e. Time Management: Presentation too long, too short, or appropriate.

The relationship to the audience is important since in real life being able to communicate effectively to get point(s) across is an essential skill to have acquired. To be able to display information using appropriate diagrams, graphs, or other such statistical knowledge is a useful technique and should be graded accordingly. Handling questions is a competency students will need to acquire in various aspects of their future studies and beyond and hence should be encouraged. Students will need to be able to manage effectively their own time in many aspects of life, whether it be for a timed examination or presentation.

The drawback with this type of assessment is that personal knowledge of a student's progress can often inhibit the successful marking of the presentation. It is advisable to have, if possible, a team of graders to provide input on various facets of the assessment structure. This will allow for a wider and much broader opinion and to achieve consistency and consensus between the presentations. Presentations are expensive in terms of assessor's time and tend to be subjective, and students do not always like the idea of standing up in front of their class and teachers. However, a well organised project plan and relaxed atmosphere in the room used for these presentations should help to overcome this. The benefits that can be obtained from this type of assessment far outweigh the disadvantages.

TEACHING STRUCTURE

Project work does not necessarily mean that all teaching in that subject ceases. Teaching often continues and some time should be set aside during each class to complete the project. During the project time the teacher becomes a facilitator of learning and aims to become an equal participant in the inquiry process. The roles of the teacher and student become more mutually supportive rather than separate. Rather than explaining the answer to a specific problem, the teacher serves as a guide and advisor. This enhances the readiness of students to learn in a self-directed way, and enables them to become progressively responsible for their own learning and development and to use their own experience as the relevant starting point for structuring learning.

If teaching continues as a project gets underway, the relevant knowledge that is required by the students to start the project should have been covered previously. If the subject matter has not been addressed previously by the teacher, problems of project selection and methods chosen to implement the solution, may be severely hampered. Careful project development planning is vital and builds upon good practice that has been established in the classroom. This planning provides a means of supporting the students in developing their projects as it enhances existing teaching. It also enables students to identify personal, educational, and vocational goals and draw up plans through which these goals can be realised.

When teaching has been completed and project work takes up the entire time there are several points, listed below, that need to be considered.

- All students should be engaged in meaningful activities. The teacher should monitor groups to prevent one member of the group doing most of the work while the rest do very little.

- Motivation of students needs to be a paramount issue, since the enthusiasm that may have prevailed at the start of a project may start to diminish over time unless monitored sufficiently by the teacher.

- The amount of time spent on a project necessitates careful planning, otherwise other lessons may suffer as a result of students over-indulging in the activities of the project to the disadvantage of other subjects or classes.

- Supervision of the projects is vital and stages of work should be observed and graded systematically, recorded and reported back to student project groups.

The above points may also relate to projects where teaching is taking place but experience has shown that these items are more prevalent when teaching has ceased. In the UK it has been recommended that project work is not left to be undertaken at the end of a course but rather an integral part over the duration of the course.

GUIDING OUR STUDENTS

Students can use the following type of outline to focus their attention on the various areas of project work to be completed. Further discussions can take place with the teacher as required during the progression of the project.

1. Group discussion of what they wish to find out and how to proceed.
2. Draft questionnaire to be produced.
3. Pilot of the questionnaire.
4. Revised questionnaire in light of the comments received for the pilot.
5. Full sample taken.
6. Analysis of results.
7. Production of report and materials used for presentation.

The use of a help factor

We must keep in mind that the problem of providing instructions for students is to avoid telling them so much that interest is lost or telling so little that they do not know what is expected of them. A possible scheme for enabling students to get a good start to a project is to make use of a "help factor." For example, consider a situation where the ADIE model is used for assessment with respective maximum grades being:

Analysis	Design	Implementation	Evaluation
10	8	14	8

For each section a help factor can be applied with suggested weightings, as shown in Figure 1.

No help				Total help
1	0.8	0.6	0.4	0

Figure 1. Help Factor

This means that if a teacher gives just a little bit of help in a particular stage (say, Analysis of Problem), the score for this stage will be 0.8 of what it should be. If a teacher gives a considerable amount of help to a student, then the help factor would be 0. Although this may mean that a student effectively receives no grade for a stage, it is still possible to attain high grades in the other sections. An example of this usage, which has been adopted by some of the independent examination boards in the UK for project grading, is shown in Figure 2.

Component	Possible Grade	Grade Given	Help factor	Adjusted Grade
Analysis	10	10	0.6	6
Design	8	6	0.4	2.4
Implementation	14	12	1.0	12
Evaluation	8	6	0.6	3.6
Total				24

Figure 2. Example of Help Factor

FEEDBACK

The assessment of projects is clearly a complex matter and there is no single agreed way of achieving this with total success. Different philosophical perspectives interact with different practical constraints to produce a variety of procedures to follow. The teacher must decide which is the most appropriate method to use under the circumstances.

The purpose of assessment is to provide information about the students who have been assessed. Feedback should be given as soon possible after the project has been graded. Feedback can be negative as well as positive since students need to know their strengths and weaknesses. If a staged assessment is used feedback can be given at the end of each stage so that students know exactly what grades they have achieved up to that point.

Incentives are often given in everyday life to encourage people to complete a certain task. As an example, Ford Motor Company récently sent out a market research questionnaire to car drivers with the incentive being a free music cassette for every completed questionnaire that was returned. In the case of students' projects, a prize of a book token could be used to induce motivation. Competition plays an ever increasing role in society and students seem to accept this fact and appear to be happy to compete for some form of reward. Nationally, in the UK, the annual statistics competition provides prizes for the best statistical projects for different age groups. This scheme has been very successful. This is not practical for all project work but, if possible, it is a method worthy of consideration.

Verbal feedback is a method by which a student can respond immediately to correct errors in the project and prevent future mistakes. Feedback provides a useful mechanism for delivering comments which may be lost in the formal grading scheme. This method can be used by teachers and students alike to create an atmosphere of encouragement and learning.

IMPLICATIONS

For any teacher undertaking to use project work within their courses there are several issues that need to be considered. These issues are summarised below:

- Whether projects are to be undertaken by individual students or groups of students.

- If group projects are used group formation needs to be decided.

- What type of project is to be used, i.e., is the project chosen by the student or set by the teacher?

- How the project is to be assessed. Students should be aware of this structure.

- Time available to complete the project should be stated, thus enabling students to organise their workload.

- Who is responsible for supervision and smooth running of the projects.

- How the project work fits in with the overall structure of the course. What teaching will take place both before commencement of the project and during the projects' time span.

- Students embarked on a project will need to be given access to sources of information and other materials that will enable them to work at an appropriate pace.

Teaching, if it continues during the project, needs to be structured so that students can get the maximum benefit from the project and the classroom teaching. If teaching has finished, then motivation of the students is a paramount factor. If students lack motivation this could lead to the project being very weak. Project work should be an active method of learning, requiring decision making and problem solving on the part of the student in "real life" situations. Where students have chosen their own projects, there is a marked interest and an enhanced motivation factor, and hence should be encouraged where possible.

If mixed ability grouping is used, the distinction between the most and least able students can be blurred, so this type of group work was found to be the most appropriate. Group project work provides an opportunity for students to use real life situations. They also help develop interpersonal relationships through understanding of the other student's point of view.

Projects that do not succeed are because students fail to start with precise aims and objectives. Since assessment systems play such an undeniable role in determining how learners and teachers spend their time, it is crucial that such systems focus on what we actually want learners to be good at. For a project to be successful, teachers should ascertain that students are able to:

- Solve novel problems, best chosen by the students themselves (see list above), but possibly devised by teachers, using facts and principles previously encountered.

- Devise an approach to investigate the chosen problem using appropriate tools and techniques.

- Keep adequate details of all work attempted and completed.

- Use appropriate procedures and formulae when calculating results.

- Communicate results obtained in the most appropriate manner.

- Discuss the results and inferences/conclusions obtained.

This model is by no means the only way of advising students doing project work. But it will give the students and teachers a framework on which to build and hence minimise the probability of a project failing. If students start with a clear problem scenario and with precise goals to achieve, and produce a detailed outline of work to be carried out, a successful project should result.

If we carefully guide our students as they develop projects, with clearly defined aims and objectives, project work allows for the integration of several previously encountered techniques. It may also be designed as a learning process in which students are faced with new concepts and unfamiliar activities. A point made by Kelly (1978) which is worth remembering is that assessment is a method of teaching, not an alternative to teaching.

Since education is our chief concern, one of our main goals must be the autonomy of the learner. Project work is a means by which this autonomy can be developed. The use of project work to create a situation in which students' learning is increasingly self-directed and self-propelled is highly recommended.

Chapter 12

Assessing Project Work by External Examiners

Peter Holmes

Purpose

Over a number of years the external examiners of a regional statistics course for 18-year-old students in schools and colleges in the United Kingdom became aware that the method of assessment was distorting the teaching and learning process, that the things being assessed were not the things that the examiners thought most important for the students to know. This chapter shows how the assessment methods were changed by adding in a compulsory project and reflects on the impact of this change on the teaching and learning of the students.

BACKGROUND

In England and Wales many 16 to 19-year-old students concentrate their studies on about three subjects, and at the end of two years of concentrated studies are assessed for what are called their Advanced Level General Certificate of Education (ALGCE). Some students will choose to specialise in mathematics, which may include some statistics, and a small number choose to specialise in statistics. Typically the syllabuses for these subjects are set by independent examining boards. The form of assessment and its implementation is also decided by these examination boards. Examination papers are set and marked by external examiners appointed by the boards. The only contact the students have with these examiners is through sending completed examination papers and other course work to them for assessment. Historically the main means of assessment, particularly in mathematical subjects, has been through three hour examination papers which the students have to answer on their own, without recourse to books or to each other or their teachers. In mathematics they may take one such paper in statistics. In statistics they will usually take two such papers. The results the students obtain in these examinations are highly influential in deciding whether, and to which, university they proceed. There is therefore a lot of pressure on them to get high marks in these examinations and other assessments made by these external boards.

From the external examiner's point of view it is clear that this scenario leads to a sort of game in which the student is under pressure to gain as many marks as possible, and this can be detrimental to encouraging a student to gain a deep understanding of the subject. Over the years examiners were becoming aware of the effect of the form and content of assessment on teaching and learning.

The challenge was to reconsider the impact of the assessment methods on students and their teachers. How could we change the forms of assessment to encourage students to gain a deep understanding of statistics, even if the student's aim was only to get as many marks as possible?

(These two aims do not necessarily coincide, and in our case they certainly did not.) How could we change the form of assessment to encourage the right sort of teaching and learning?

For reasons given below, it was decided to make the students carry out a compulsory project as part of their course and to have this assessed. Also, for reasons given below, it was decided that the students should be able to choose the topic of their project. Since the ALGCE had a reputation of being an objective assessment of students' abilities any form of assessment had to be monitored and standardised, at least in the final analysis, by external independent examiners. The teachers or tutors of the students were not to be the final arbitrators of standard since it was considered that this was too subjective and arbitrary.

Since this was a new departure for students of statistics at this level, a number of questions were raised and had to be answered.

- Having decided on free choice projects, how could the external examiners assess the different projects?
- If any internal assessment were to be carried out by the teachers or tutors of the students, how could it be taken into consideration?
- Since the number of students entering these examinations was too large for all the projects to be assessed by a single external examiner, how could the marks of these external examiners be standardised?
- How do you cope with the limitations of external assessment when you can't see or interview the student?
- Since this was a new departure, what was the effect on students' learning?
- What was being assessed that previously was not being assessed?
- What showed up in this assessment that we had not previously known or had not been able to assess?
- How successful was incorporating project work into the assessment in improving the statistical understanding of the student?

I shall attempt to answer these questions based on my several years' experience as chairman of examiners and chief examiner for one of the ALGCE syllabuses in statistics that went down this road of broadening the assessment.

EXTERNAL ASSESSMENT OF COURSES IN STATISTICS

Curricular goals and principles

The position before the change

There have been ALGCE syllabuses in probability and statistics for over 30 years in England and Wales. The goals and principles of assessment reflected their historical development and were not properly geared to the goals of a current statistical education. Part of the difficulty was that they had always been seen as part of a *mathematics* qualification. Earlier syllabuses had sections in pure mathematics and in applied mathematics (interpreted as mechanics). Since other subjects, such as economics, biology, geography, psychology, and social sciences were becoming more quantitative, there was a strong argument for replacing the applied mathematics section with a section on probability and statistics. It was thought that this would make the syllabuses more useful for students whose main interest was in these newer quantitative subjects. So new syllabuses were drawn up—but since they were part of a mathematics course they were drawn up by mathematicians. They were drawn up so that they could be seen as mathematically respectable, and so they concentrated on mathematical statistics. The three-hour end of course exam was taken as the appropriate assessment method, since this was taken for granted as being an acceptable means of assessing mathematics.

Several side effects followed from these basic decisions:

1. Questions in the examination would often concentrate on mathematical principles rather than statistical insight. For example, the student might be asked:

A probability density function has the form $f(x) = ax^2+bx+c$ with $0 \leq x < 1$.
It has mean 0.6 and variance 0.1. Find a, b and c.

This requires no knowledge of statistics and only elementary knowledge of probability. It is essentially an exercise in solving three equations in three unknowns.

2. Questions which did try to explore some practical implications of statistical conclusions (such as asking the student to explain the practical implications of a given result being statistically significant) were notoriously badly answered, and also tended to carry few marks in the assessment. In general the form of the three-hour examination meant that very little emphasis was placed on the practical implication of the final result; indeed, most questions were based on false data invented for the question, so no true practical conclusions could be drawn in any case.

3. Because they were drawn up by mathematicians, the syllabuses were less useful than they might have been to the very students it was hoped would take the subjects. They did not link the statistics with the use being made in the other quantitative subjects.

4. A most important drawback was that the syllabuses, and particularly the assessment, did not encourage the students or their teachers to consider the statistical expertise required to carry out a complete statistical investigation. They did not encourage practical statistics or a consideration of real data and real problems. They were not developing true statistical expertise.

Reflecting on the position before the change

It was in considering these effects that we were led to ask whether what we were doing was best—and decided that it wasn't. We needed to sharpen up our aims in these courses and try to gear the assessment to reinforce the aims. It was clear that we needed to consider again what it meant to be a statistician at this level; what are the appropriate levels of statistical expertise we are trying to develop?

What is statistics? It is not a subset of mathematics, nor is it a set of techniques. Any definition must be broad enough to include all that statisticians do—applied statisticians in their different fields as well as academic statisticians. Statistics is a practical subject devoted to obtaining and processing data with a view to making inferences which often extend beyond the data. The statistician is involved with helping people make decisions in the face of uncertainty.

Including projects

Our reflections on these lines led us to decide that a compulsory project as part of the assessment would be a good thing. Since we wanted the students to be able to link this project with their own interests or with one of their other academic subjects, we decided to make the topic of the project open to the student to decide. We felt that through projects we could encourage a more rounded approach to statistics. There were also important things we could assess through the project work that we could not assess through the traditional three-hour exam. By a project we meant a piece of work undertaken by the student that would take about 40 hours of work, plus any time needed for data collection, that would start with defining a problem, collecting the appropriate data, analysing the data and drawing appropriate inferences. All this was to be presented in a written project report of about 15 pages.

The positive effects of projects

The team working on this change in assessment decided that there were many positive gains in requiring students to carry out a practical statistical project.

- Projects *put statistics in a context*, and make them more relevant. If the data arise in considering a particular problem, they have meaning. We do not have the problems of invented data; the data are throwing light on the situation. They are asking to be interpreted.

- Projects *give more motivation*. This is particularly so if the topics are chosen by the students themselves. They then do have a built-in desire to get insight into their problem. Sometimes the conclusions can be of real practical importance. I remember reading one project submitted by a student who had a weekend job at a local pharmacy. He was going to use data from this pharmacy as part of his project. The week before he was due to collect the data the pharmacist said that he was no longer required. The student persuaded the pharmacist to let him work for two more weeks (unpaid) whilst he and his friend collected data. They collected data on the number of customers, waiting times, the number of people who looked into the store but did not come in because it was too full, the amount of money collected in payment by the student, and so on. They then reworked these data on the assumption that the student had not been there, making a few other assumptions (usually in their own favour). At the end the report showed that the pharmacist would be losing money if he did not employ the student. As a result of showing the pharmacist the report the student got his job back. The student found real motivation in doing that project, and also completed a very good report.

- Projects *give a greater feel for real data*; their accuracy or otherwise; variability; reliability of conclusions; measurability. It is in collecting the data that you realise the decisions that have to be made, the way that some people are not necessarily answering the questionnaires truthfully, the possibility of bias in some of the measurements, and so on. It is this sort of feel for data that is most important when it comes to analysis and interpretation of the data.

- Projects *emphasise the application of statistics and its usefulness.* This can be a side effect in a larger group of seeing many students doing projects in many different areas. It is a good counterbalance to the impression that they can get from textbooks and from traditional examination questions.

- Projects show that *statistics is not solely mathematics.*

Projects or practical work?

An alternative to including project work as part of the assessment was to encourage practical work. The difference between practicals and projects can be summarised as follows.

A practical introduces or reinforces some particular theory; a project links a number of topics. Also, a practical has clear cut and limited objectives; a topic is investigative and more open ended.

In a practical the teacher defines the problem, decides the model, and decides the techniques; in a project the student defines the problem, chooses the appropriate model or models and decides the techniques. A project also requires a more substantial report to be written, while practicals are more easy to control than projects but do not develop global skills.

So, although practicals have a very important role, they do not develop the skills we wanted to encourage in the same way as do projects. We therefore chose to incorporate projects into the assessment process rather than just try to encourage more practical work. In fact we do try to encourage practical work as a precursor to projects, but that is not part of the assessment process, so is not necessarily done.

Anderson and Loynes (1987) give other good reasons for doing project work as part of a statistics course. They are writing from a university undergraduate and postgraduate perspective, but their points are still valid. It is interesting to see that they talk of skills that would not necessarily be considered statistical, but are part of good learning. They write of general non-technical abilities, such as working with others and communicating; general, partly technical, abilities, such as appreciating the ethics of a statistician and determining the aims of an investigation; abilities depending on technical skills, such as recognising which techniques are valid; and abilities depending on technical judgement, such as translating a real problem into statistical form. It should be obvious that it would be very difficult to assess these skills by the traditional three-hour examination.

Examples of project titles

One of the main purposes of including projects was to get the students to own the data, so we allowed them to choose their own topics. This did pose problems for assessment, and these are considered below, but it meant that a wide variety of topics was chosen that reflected the wide interests of the students.

As an example, here are some of the topics that students chose in the first year of the new statistics course:

- Standing in line for school dinners
- Pupil preference for different chalk and board colour combinations
- The effect of "rounding" on regression lines
- Effect of soil conditions on the distribution of wild plants
- Human reaction times and learning ability
- Quality control of blood testing
- Consumption of gas in the UK
- Connection between rainfall and land height
- Predicting football results
- A game of golf
- Number of assistants required at a local shop
- Petrol sales at self-service and attended garages
- Mistakes in newspapers
- The effect of an overhaul on a canning machine
- Delays in flight arrival times at Manchester International Airport
- Analysing crime statistics
- A dental survey of children's teeth in three schools

In general the better projects were those that were more tightly focused and that drew on the real interests of the student. Projects which tried to draw on large data sets, or were too ambitious, were less successful.

Principles which should guide assessment

The experience of examiners in the pure mathematics and statistics courses was that teachers were training their students to answer the questions being set in the three-hour examinations. Much time was being spent on teaching for this purpose, and it was distorting what the course was meant to be about. In particular it was not encouraging teaching that would lead to good statistical insight. I do not intend these comments to be seen as derogatory about teachers. If we as assessors imply that some things are important by the way we design our assessment, then it is reasonable for a teacher to try to enable the students to get as high an assessment mark as possible. The whole exercise becomes a game to get the most marks out of an examiner. This is particularly the case when, as here, the marks can be crucial for the student's future career.

Many people have written more recently about what are good principles for assessment. The two major ones we took on board at the time of changing the course under discussion were (a) that the assessment should encourage good teaching and learning and (b) that the assessment should be objective and consistent.

We wanted to frame our assessment procedures to encourage good teaching and learning. Our assessments should seek to ensure that the content of the courses is right and that they develop the insights and abilities that are important. This includes process and global skills as well as particular techniques and topics. If we think that co-operation with others is important we must include that in the assessment, even at the expense of having difficulty in evaluating an individual student's contribution to a joint effort. In technical terms this means that we may have to trade off some reliability against having greater validity. If we think communication skills are important for a statistics course at this level, then we should include them in our assessment.

The second principle in external assessment of the sort we were doing is that it should be objective and consistent. A mark scheme should be applicable to all legitimate types of project. If different markers are used they should come up with essentially the same marks.

Although the nature of the final assessment of the project was summative in that a mark or grade was given for the record, we wanted to encourage the project to be used as part of the teaching. So we involved the teachers with the project work so that they could do formative evaluation to help students learn before making the final evaluative assessment. Since the teachers were closely involved with the students doing the work, they actually had deeper insight into the nature of the student's understanding than was possible for the external examiners. So it was made a principle that the teachers do their own assessment and send these assessments and additional comments to the external examiners for them to use in their assessments.

Pragmatic examples using target knowledge or curricular goals

What is an acceptable project topic?

After deciding that a major project should be part of the assessment, we had to decide what sort of a project we would accept and hence encourage the students to do. After discussions with statisticians and teachers as well as with other examiners, we decided that anything which was a practical investigation using data would be accepted. It did have to be practical and it did have to use data. We wanted to allow students to use either primary or secondary data, or both, as appropriate. We did want them to do more than a trivial analysis of these data. We wanted the project to use some of the techniques that were part of the general syllabus content. This meant that we had to write our assessment scheme in sufficiently general a way so that all such projects would be included, and so that it encouraged a good use of the statistical techniques.

This decision did rule out certain types of projects that in other circumstances may have been quite legitimate. For example, historical projects such as *The rise and use of the Least Squares Principle in statistics,* and basically theoretical projects such as *The development of the theory of Markov Chains in probability* would not meet the guidelines. In fact, in the first year, the Markov Chain example was submitted. In its own terms it was a good project, but the teacher responsible had given poor advice to the student. To prevent the student from being unduly penalised, we, the external examiners, kept a close eye on the other papers done by this candidate so that she would not be unduly penalised. As it turned out, her examination papers were excellent and, even retaining the low mark for her project, she gained the highest grade on the assessment as a whole.

How much weight should be given to the project?

Another decision we had to make was how much of the total assessment mark would be given to project work. Too little and students and their teachers would infer that it was not all that important; too much and there would be political pressure (in the general sense) because of the perceived lack of objectivity in marking projects and in the possibility of students receiving help with their projects. Eventually we settled for 20% to the project and 80% to the final examination papers as being politically acceptable. In terms of the effect on teaching and learning I think it would be better to move towards a 30:70 split and require more effort to be put into the project.

Marking the projects

The assessment had to be sufficiently general to cover all acceptable projects (as defined above) and had to be sufficiently precise so that teachers could carry out their own initial assessment and so that the team of external assessors could use it consistently.

The scheme set out to assess four qualities:

A The ability to select an appropriate probability model and set of statistical techniques for the investigation of the topic, and to design a plan of procedure;
B The ability to carry out a practical investigation in accordance with a plan of procedure;
C The ability to present the results of practical investigation in a valid and informative manner; and
D The ability to draw conclusions from practical experience and evidence.

The total marks were out of 40, divided A(8), B(9), C(10), and D(13).

To allow for the different levels of technique used in the different projects there were several general principles, such as:

- Mark generously for a difficult project involving difficult techniques, difficulty in collecting data, etc.
- Make appropriate allowance for the amount of direct help and guidance given (a form had to be filled in by the internal assessor and sent to the external examiner describing this).
- Discount work which is irrelevant to the project.

More detailed general headings were given under "A" to "D" to help with the assessment and allocate part marks.

For example, in "A" you were told to look for a description of aims and objectives, a description of the plan and design, the nature of the plan and its appropriateness to the aims and objectives, the probability models and statistical techniques used and their appropriateness. Later this was expanded for the benefit of the teachers who were doing the initial assessment. This included such statements as credit should be given for suitability of models, number of different models, and difficulty of the models used appropriate to the data. The description of the plan should indicate that some thought was given to such things as the data needed, how to collect the data, and how to use the data.

In "D" you were told to give credit for carrying out the analysis clearly and accurately, drawing valid statistical and practical conclusions from the data, discussing other possibilities and limitations of the project, and summarising the main points succinctly.

There were similar headings for the other two sections. External assessment followed internal assessment using the same scheme. The teachers made their assessment and completed a form to say not only what was their assessment but also what help the student had received. They had to complete a declaration that it was the student's own work (except where due credit had been given in any references) and any other relevant comments.

The effect of the change

It soon became clear that the students were coming to the end of their courses with very different understandings and expertise than previously. Apart from the specific things we were able to identify in the projects, which are described below, there was a real difference in their answers to questions on the three-hour examination. In particular, the questions that had a request for a practical interpretation at the end were now answered much more sensibly. Whereas in the past we may have had an answer such as, "The result being significant shows that the data are significant," we were much more likely now to get comments relating to the context of the question, and even whether the underlying assumptions (say of the normal distribution or of independence) were correct.

Including project work meant that we could assess qualities in the student which would be very difficult in the traditional three-hour written examination. Even an oral examination would not have been as effective, and cost ruled this out anyway. Project work enables us to assess the student's ability to carry out a complete investigation from start to finish and to communicate the results. It reveals the ability to plan, work co-operatively, use limited resources effectively, adapt a plan as circumstances demand and complete it on schedule, and so on.

Pupils' difficulties as indicated by project work

After the scheme had been running a few years, we realised that we had found out much that we would otherwise not have known. We saw difficulties that the students were having that we had not known before. There were difficulties with the design of the project, with the analysis of the data, and with the written reports.

The difficulties in *project design* were associated with learning how to focus in on a sufficiently tight project that could be carried out in the time and with the resources available. We would find projects that did not succeed because they started with imprecise aims. At one extreme this could be the sort of project that worked along the lines of "let me find out all the data that I can that is at all relevant to my particular interest and see what I can deduce from these data." Such projects were not sufficiently well focused.

Other projects fell short because they were too limited in scope, or had repetitious designs; for example, a series of t-tests all of which were essentially the same.

Several projects could easily have been improved if there had been an appreciation of elementary design of experiments. For example: avoid confounding, use matched pairs to lower variability, use randomisation to eliminate other possible effects. An elementary consideration of the effect of sample size on the significance of results was also sometimes missing. Sometimes the samples taken were so small that it was impossible to detect any effect.

There were some surprising difficulties associated with the *analysis of the data*. It was not the purpose of the project assessment to look for details in the carrying out of the analysis. It would have been an impossible task for the examiners to check all the calculations. Even so, we did look to see that the summary figures and calculations bore some resemblance to the data. We expected most errors to be picked up during the normal consultation process between the student and the teacher whilst the project was being done. Nevertheless it was surprising to find some gross errors.

There were errors in diagrams. For example, a fitted regression line would pass above all the points on a scatter diagram. There were also errors in the arithmetic. For example, the mean of a set of data was greater than the maximum value.

There was often a cavalier approach to data. This sometimes came from not knowing how to deal with outliers. Sometimes, genuinely, outlying values were dropped from the analysis because there was some good reason for thinking they were in error rather than just examples of high variability. Occasionally this process was taken to extremes with students automatically not using the highest and lowest values.

A continual problem was with students using significance rather than estimation. This was partly a function of the syllabus which emphasised hypothesis testing, but estimation and confidence intervals were also included and often this would have been a more informative approach to the data.

There was an occasional confusion between regression and correlation when students were not sure whether it was the strength of association or a prediction that was required.

The main difficulty that students had with their *reports* was a failure to draw valid conclusions based on statistical analysis, and prejudice showed through. One example was a study done on driving abilities of male and female students which, if anything, showed that females were better drivers than males. The student, a male, drew the conclusion that although the data seemed to show that females were better than males, everyone knew males were better.

Some students found it difficult to get the right balance between theory, detailed calculation, and summary in the report. The report has to show that the student has understood what is being done, but does not have to include large amounts of theory or repeated detailed calculation.

In spite of the above negative comments, the large majority of reports showed a much deeper understanding and ability to carry out a full statistical investigation and communicate the results than had previously been the case.

One area of concern was communicating with teachers. Because many teachers were involved, spread over a large geographical area, it was not possible to do other than have a couple of central meetings. Apart from these, communication was by printed material. The projects and the final marks awarded were returned to the teacher. This did help the teachers, over the years, to bring their assessment standards into line and to get experience in which subjects were and which were not likely to be good projects.

Realistic limitations and difficulties

The major difficulty was in devising the assessment to reinforce the learning principles. This has largely been covered above. There were other limitations and difficulties.

Initially there were not large entries to the examination—in the low hundreds. This meant that every project report could be collected in and re-assessed by the team of external markers. This small team standardised their marking by cross marking several projects and comparing their marks. Given the diverse subject matter they became quite consistent at this. This is important because such consistency is necessary if the assessment is to be reliable. A high level of reliability was expected for this examination as a whole. We all felt that the price of a little less reliability in order to gain greater validity (we were assessing the things we wanted to assess) was well worth paying.

Teachers seemed to be good at getting a reasonable rank order of the projects submitted from their school or college. Initially they were understandably less good at getting the absolute standard right. This was helped in succeeding years by the return of the projects from previous years and also by publishing a more detailed description of how marks should be allocated.

When large entries were received there was a plan to standardise by sampling. There was never any plan to standardise by reference to the results of the three-hour examinations since this would give extra weight to the examination, which did not measure the same things as the project. We felt that if we standardised against the examinations the students would concentrate on getting good marks in the examination at the expense of doing a good project, and we would be back where we started with bad assessment encouraging bad learning.

There were some difficulties involving how properly to incorporate teacher comment. The external examiners had to rely completely on the written reports from the teachers, which were fairly brief and variable in quality. In fact, only minor changes of marking were made if we received comments indicating large amounts of help being given. Our experience was that teachers were so anxious about giving too much help that they erred in the other direction.

Initially the teachers found their dual role difficult; they were both teachers and assessors. We tried to encourage them to see themselves as consultants, which was different from teaching. Their role was in asking questions rather than in giving answers. In practice the teacher's role changed throughout the course. Initially they were helping the students decide on a topic. They were helping them identify an area of interest, to see what questions could legitimately be asked, what data might be collected, and they were giving some sort of idea of the sort of answer the students might expect to obtain. Sometimes this took them out of a field of knowledge where they had any expertise. They had to develop the facility of asking probing questions and learning about the context from the student or other colleagues.

During the project their role was still as a consultant, and as an encourager to the student. They could also point out how the techniques they were teaching might be used in different projects, and use some projects as examples in their teaching. Before the final assessment the better students saw their teachers. At this time a teacher might spot major blunders and ask a question such as, "Are you sure that is right?" One difficulty was with students who did not keep in touch with their teachers, worked on their own, and submitted their projects at the last minute. Here there was no opportunity for the project to be corrected before assessment.

There is no doubt from our experience that the nature and form of assessment on the way the subject is taught and learned, and on what is seen by the students, is important. Including the project as part of the assessment did have many of the effects that we wanted.

IMPLICATIONS

It is clear from the effect of the changes we made that assessment does affect learning, attitudes, priorities, and the view that students have of the subject. We can not afford to continue with inappropriate assessment no matter how long the tradition behind us. Positively, if we orient our assessment towards assessing what is important, then this does change our students' attitudes to what is important (it also has a similar effect on some of the teachers). Rather than carry on as we have always done we need to think carefully about the purposes of our course and gear our marks to this purpose rather than vice versa. We must try not to let the assessment be seen as the most important aspect of the course.

If you are carrying out a change such as we did—giving a greater emphasis on project work and putting more responsibility on the student to choose a topic and on the teacher to supervise—

then be prepared for opposition. It will come from people who will argue that the assessment is losing reliability and is less objective and less under the examiners' control. The response to this is that the total assessment is becoming more valid, we are assessing what ought to be assessed rather than what is easily assessed, and to some extent there has to be a payoff of reduced reliability to get greater validity; however, every effort has to be made to make the new form of assessment as reliable as possible. There will also be opposition from those teachers who do not want change, for they have become used to what they are doing. These teachers need to be persuaded of the correctness of what is being done and every effort made to quell anxiety and fear and to equip them for their new role in the whole process. In our case, if we had not done this the teachers would have transferred their students to other courses of a more traditional type. We provided meetings at which teachers could hear about and discuss the changes, ideas for possible projects, the reasons behind our changes—the change of attitudes we hoped to encourage, examples of projects, marks and other information. Once the new scheme had been started other in-service providers gave courses to help teachers in different ways.

Don't underestimate the effort needed to make such changes, but neither underestimate the harm that can be done by wrongly focused assessment and the good that can be done by well focused assessment. It may take a lot of time and effort, but the gain is worth it.

Chapter 13

Portfolio Assessment in Graduate Level Statistics Courses

Carolyn M. Keeler

Purpose

In this chapter, the process of developing a form of alternative assessment, the portfolio, will be described. The portfolio is a purposeful collection of student work that exhibits the student's efforts, progress and achievements over time. Portfolio development supports the assessment of long-term projects, encourages student-initiated revision and provides a context for presentation, guidance, and critique. The purpose of portfolio development is the same no matter the course or age of the students, to display the products of instruction in a way which challenges teachers and students to focus on meaningful outcomes. The context in which the use of portfolios is described here is a graduate level statistics course where an additional purpose is to provide students with an organized reference on statistical programming, analysis, and interpretation. However, the process used in developing portfolios and the important issues surrounding portfolio assessment can easily be generalized to different educational levels and subject areas. Some of the questions addressed in this chapter are 1) What is the underlying belief concerning knowledge construction which guides portfolio assessment? 2) How do you develop and use portfolios? 3) What does a portfolio look like? and 4) What are the major considerations in deciding to use portfolio assessment?

INTRODUCTION

Teachers are beginning to recognize that traditional tests are limited by easy scoring formats and the testing of easily retrievable fragments. The skills that result from students passively absorbing information and taking multiple choice or true-false tests may have little or no relevance outside of the classroom because they are not transferred to everyday problems. Research from the field of psychology indicates that learning does not occur by passive absorption alone. In many situations, people learn using prior knowledge, assimilating new information, and constructing their own meanings (Resnick, 1987). This constructive, active view of the learning process is reflected in the way much of mathematics is taught. In rewriting curriculum to allow for the construction of meaning by the learner and the application of learning to solve real-life problems, the dilemma is that we must also change the way we assess curricular obtainment. Assessment has the potential to teach and gives the student opportunities to develop complex understandings.

Good assessment enables us to accurately characterize students' functioning and performance in order to make sound decisions to improve instruction. Portfolio assessment is an example of assessment integrated with instruction and is one alternative to one-dimensional testing. The Encyclopedia of Educational Evaluation (Anderson, Ball, & Murphy, 1975) defines assessment as a process built around multiple indicators of performance:

Assessment, as opposed to simple one-dimensional measurement, is frequently described as multitrait-multimethod; that is, it focuses upon a number of variables judged to be important and utilizes a number of techniques to assay them...Its techniques may also be multisource...and/or multijudge (p. 27).

Portfolios are by their nature multisource. Demonstration of what students know and are able to do takes many different forms in the portfolio. Students, teachers, or students and teachers together may choose what example to include of the activities completed for each different task. Although some tasks may have been handed in previously and revised, the student writes the reasons for including the example and a critique of what was learned and demonstrated (or not). This is both a learning and assessment strategy: students learn through the critique of their work and demonstrate the level of understanding they now have of the concept. This in turn informs the teacher of the need for further clarification or reteaching.

With some variation, the process for developing portfolio assessment is the same as that used by developers of performance measurements no matter what their nature. (See also chapter 3 in this regard.) This process is offered here in seven steps. It is generally accepted that the process should conclude with the revision of both instructional and assessment tasks.

Step 1   Delineate the knowledge and performance and process skills students are to develop.
Step 2   Develop assessment tasks which require students to demonstrate the knowledge and skills delineated.
Step 3   Specify the criteria for judging student performance on each task.
Step 4   Seek evidence of validity in measuring the knowledge and skills delineated in the course content through expert panel or item analysis.
Step 5   Develop a reliable rating process or rubric for each task.
Step 6   Provide feedback to students.
Step 7   Refine assessment and improve curriculum and instructional tasks.

The chapter is organized under four headings, 1) the constructivist framework, 2) portfolio development, 3) portfolio sketch, and 4) implications, issues and concerns. The section on the constructivist framework describes this theoretical basis for instruction. Students are asked to construct their own understanding of the subject matter through active involvement in the learning environment using authentic learning tasks and application of the knowledge gained to real-life predictable and unpredictable problems through outside projects. The portfolio development section includes the purpose of portfolio assessment, development, creating and critiquing assessment tasks, and establishing the scoring criteria.

In the portfolio sketch, one example of an application of the development process is offered. In the example, portfolio assessment is used to demonstrate mastery of computer software programming and the application of statistical processes to projects and exams. A reflective journal is included to emphasize the importance of student reflection on the learning strategies used to acquire the knowledge and application of knowledge to course products. Finally, the scoring procedure is described.

The final section covers implications for use of the portfolio and some issues and concerns which emerge as a result.

THE CONSTRUCTIVIST FRAMEWORK

In the early 60' s following the launch of Sputnik in 1957, the focus of education in the United States became less on liberal arts and more on the scientific and theoretical base of knowledge. As a result, a philosophy called objectivism pervaded the methodology of teachers. It holds that knowledge exists independently of the mind for transfer into a student as if into an empty vessel. In the field of statistics this would be realized in teaching emphasizing the rote memorization of formulas. In the 90' s, the constructivist framework for education is becoming quite different. Rather than transmitting knowledge through texts and lectures and asking students to memorize facts, the model for the teaching-learning act becomes one of transaction, where meaning is made through a significant transaction with the information (Pulaski, 1980; von Glasersfeld, 1981; Davis, Maher, & Noddings; 1990; Schifter & Fosnot, 1992). The main idea is that learners must construct their own understandings rather than passively absorbing or copying the understandings of the instructor. In my practice, this constructivist view of learning is not only the theoretical basis for instruction but is also a principal goal of instruction. When learning is an active process the student demonstrates understanding through such products as reports, speeches, and models. When these products become part of a portfolio, they may be extended, revised, or reflected upon by the learner and thereby become one form of assessment used to determine the student's progress.

As students actively engage in problem situations, they build understandings which are an extension of and then become a part of current knowledge. Learning is the result of an action in which students participate, not the inevitable product of encountering materials. As part of this meaning-making, opportunities are provided for students to pause and reflect upon what they have been doing. The instructor can assist students through simple devices such as pausing and asking students to recount to a neighbor the problem solving process in which they have engaged, leading group discussions of the thought process of individual students, or having students reflect in a journal regarding the development of their own understanding.

The radical constructivist perspective suggests that the individuals' understandings are built from their unique web of prior concepts and their current subjective experience and therefore are idiosyncratic (Schifter & Simon, 1992). It is imperative then that the assessment of this unique understanding include problem solving opportunities which parallel learning activities and opportunities for reflective expression describing the individuals' changing conceptions. For this reason, many of the projects assigned as coursework are extended and reflected upon as part of the portfolio in the form of a reflective journal.

PORTFOLIO DEVELOPMENT

The purpose of portfolio assessment

Portfolios are purposeful collections of student work that are reviewed against criteria in order to judge an individual student or a program, or both. The products on display in the portfolio focus instruction for both the teacher and the student on critical learnings. The portfolio of itself does not constitute the assessment. The "assessment" in portfolio assessment only exists when three important things are communicated to the students, 1) the assessment purpose, 2) the criteria or methods for determining what is put into the portfolio, and 3) the evaluation criteria for scoring the pieces or the collection. A portfolio in statistics education may include statistical tests and interpretation of results, projects and exams which have been corrected, and a reflective journal (see Portfolio Sketch section later in this chapter).

The portfolio has a meta-structure allowing for both traditional and alternative sources of data or multiple indicators of the same outcomes and therefore offers a more complete picture of achievement. Because it is longitudinal, it has the advantage of putting one assessment or project into perspective and documenting growth over time. It also documents what concepts and models have been acquired through the course. Portfolios should include pieces representing student progress, student reflection about their work, and evaluation criteria (Herman, Aschbacher, & Winters, 1992). For evaluation purposes, the portfolio can do what traditional assessment in most cases does not do, provide direct evidence for evaluating the student's progress over time, at mastering the essential concepts and techniques of the course. Many methods of assessment included in other chapters can yield usable results; some of these are concept mapping (Chapter 8), group work (Chapter 10), performance assessments (Chapters 3, 11, and 12), and model-building (Chapter 6).

Development

The process of portfolio development is dependent on the identification of the most important goals of instruction. In a graduate statistics course, both research models and statistical procedures are taught. Thus, students should be able to 1) choose the appropriate model and 2) formulate the research question in a way that allows them to 3) appropriately choose sampling and data collection procedures. Students should 4) know the appropriate statistical test to apply to the data and 5) how to calculate the test statistic and/or use a statistical package to computerize the analysis. Students should be able to 6) interpret the statistical output from the computer program and 7) draw conclusions from the results. (See chapter 1 for further discussion of common instructional goals in statistics education.)

Once you know what you want the students to know and be able to do at the completion of a course of study, the next step is to develop learning tasks which involve them in the discovery and application of this knowledge. Statistical solutions are not only a function of knowledge but also of metacognitive skills. It is important to allow time for self-monitoring, self-critique, and self-regulatory processes. Emphasis should be placed on making connections, understanding concepts, learning to reason statistically and developing the ability to communicate statistics effectively. The development of activities and projects which facilitate the student's ability to create personal meaning from new information and discover key concepts is the goal. Learning isn't necessarily a linear progression of discrete skills.

Authentic activities and problems to solve must be developed by which students will 1) represent problem situations verbally and symbolically, 2) integrate statistical concepts and procedures, 3) apply strategies to solve problems, 4) formulate questions and analyze problem situations, and 5) verify and interpret results. Through their discovery of these processes they begin to internalize the skills and knowledge which make the next activity easier. You cannot assess performance unless you teach performance.

Once goals and learning processes are articulated, appropriate assessment tasks may be developed to match the active, inquiry-based learning process. Cognitive learning theory encourages linking instruction with assessment yet tells us that there is great variety in learning styles, approaches to problem solving, memory, developmental paces and aptitudes. Therefore, choices should be provided in task assignments and there should be ample time for project development, revision, and rethinking (Herman, Aschbacher, & Winters, 1992). Assessment tasks should target instructional aims and allow students to demonstrate their understanding and reflect on their progress. Just as engaging students in real-world tasks is motivating, it aids in the transferability of skills and knowledge.

Alignment between the curriculum and the assessment occurs in two ways, content and context. Content alignment is necessary for assessment validity. Context alignment occurs when the assessment protocol or scenario is the same as the one used to teach the curriculum (English, 1992). Context alignment is particularly important in statistics education. When students are given traditional tests in statistics there is a tendency for recall level knowledge and memorization of formulas to dominate learning rather than the application of statistical procedures to real-world problems. Portfolio assessment which is aligned with authentic learning tasks in class can be developed which help to transfer and extend knowledge by reinforcing the application of statistical procedures and the interpretation of results.

When students solve problems, make decisions, construct arguments, defend their answers, or collaborate with others, they are demonstrating the processes involved in the discipline while gaining understanding and developing a product similar to a professional in the field (Resnick & Klopfer, 1989). Modern curriculum theorists believe that engaging students in the processes of the discipline is a powerful learning strategy (Baker, Freeman, & Clayton, 1991; Bransford & Vye, 1989; Resnick & Klopfer, 1989). For example, the Content Assessment Prototype in history, developed by Baker and colleagues (1992) at CRESST, engages students in the authentic tasks of historians. In our case, students develop a "working knowledge", one which enables them to use the tools for making, using, and communicating information through research.

Three more factors must be considered in designing a portfolio, the purpose, context and design. The purpose affects the content of the portfolio and so must be decided first. This includes what you want to show with the portfolio, i.e., mastery of concepts, understanding of the processes by which knowledge is constructed, attitudes towards statistics. Purpose also includes how the portfolio will be used, i.e., accountability, student self-analysis, program evaluation. Context includes such things as students' characteristics, activities which occur during the time the portfolio is in production and the uses of the completed portfolio. Design covers such considerations as what will count as evidence, how much evidence is needed, how it will be presented, who will decide what evidence to include, and the evaluation or scoring criteria. What is included also depends on who will view the completed portfolio. Known as stakeholders, the teacher, student, and in the context of K-12 education, parents and administrators, may be included as audiences of the portfolio.

Creating and critiquing assessment tasks

Unique to the creation of the portfolio as a form of alternative assessment, is the involvement of students in deciding what to include. The students should know "up front" how the portfolio will be weighted in the final grade and what assessment tasks, course projects, and exams are assigned as a part of the portfolio. Teacher and student should collaborate to decide what other things to include as representative of their learning and performance. Arter and Spandel (1992) summarize the kinds of questions teachers should keep in mind when creating portfolios:

- Is the work included in the portfolio representative of what students can really do?

- Do the portfolio pieces represent coached work? Independent work? Group work? Is the amount of support the student received indicated?
- Do the evaluation criteria for each piece and the portfolio as a whole represent the most relevant or useful dimensions of student work?
- Is there alignment between the course goals, the learning tasks and the assessment tasks? For example, does the task require multi-faceted cognitive skills if this is a course goal?
- Do tasks or some parts of them require extraneous abilities?

Before you use assessment tasks with students, it is a good idea to allow a period of time to lapse following development and then review the task to see if it is sound. Another review of the task should be made after initial use and scoring. Often problems with a task surface during use which were impossible to anticipate in development. The determination of the validity of the task and the reliability of the scoring rubric is simplified if you are the instructor of the course and all of the tasks will be scored by you. The teacher's knowledge of what the instructional objective is and therefore what should be assessed guarantees some degree of validity. If you are the only one using the scoring rubric, then you must be careful to be consistent between portfolios but inter-rater reliability becomes irrelevant. If you are developing tasks and scoring rubrics to be used by more than one instructor, the task should be written collaboratively and the first time the task is assigned both people should score the task and compare ratings. Some questions by which to review the tasks you have developed include:

- Can tasks provide reliable measurements? For example, is there inter-rater agreement and between-task generalizability?
- Are they sensitive to students? For example, do tasks, scoring processes, etc. avoid potential bias?
- Do all students have access to the opportunities and instruction which will prepare them to perform well?
- Are the tasks clearly delineated?
- Do they provide achievement information that differs from traditional measures? For example, does the assessment engage students in authentic or real-world tasks which are worth doing?
- Do the tasks offer different ways of assessing the same concept development or provide choices to students of tasks to complete?

Establishing the scoring criteria

Assessments should be designed to improve performance not just measure it. Feedback to students must provide direct, usable information on their performance and concrete examples of the difference between their current performance and the expected performance. Where a single score on a test is unuseful, coded information from the teacher, a scoring criteria or rubric provides detailed feedback of performance in the analytic areas.

Students should know in advance the criteria for judging student performance. The rubric includes a description of the dimensions on which tasks will be scored, the scale which will be used in rating performance, and a description of the characteristics of each level of the scale. The scoring criteria should be handed out and discussed with the students. This could occur at the beginning of the course if a generic rubric is used for scoring all tasks or at the time the task is assigned. Discussing the scoring criteria with students helps them to understand the meaning of the scale, what is expected to receive the highest rating, and the importance of the task to their learning of the discipline.

In choosing dimensions for scoring an assessment task, focus on the intent in terms of instructional goals and the performance observable through the task. Two questions which help in choosing dimensions are 1) What are the criteria for selecting the samples that go into the portfolio and 2) What are the criteria for judging the quality of the samples? For example, if communication of statistical results is an important criteria, then it should be developed as a scoring dimension and its characteristics described. In a scoring rubric, there will be as many descriptions of the characteristics as there are points on the scoring scale. If you are rating the piece on a three point scale of excellent, acceptable, or poor, then in our example there would be three descriptions of the characteristics of communication (see Figure 1 at the end of this chapter for an example).

It is also helpful if you share examples of student work which exemplify the criteria for judging performance. The first time you use the assessment task this may not be possible but it is then that you can ask students if you might retain a copy of their work to serve as an exemplar. Over the years you may find it helpful to collect examples of both excellent and poor responses to share with your students. All exemplars should of course be used anonymously.

PORTFOLIO SKETCH

Introduction

In this section, the background information given thus far in the chapter is used to form a portfolio for a specific class. The portfolio sketch is an example of how a portfolio might be developed for a graduate level statistics curriculum. The sections match the important instructional targets or outcomes of the class. The computer programs and correct interpretation of results demonstrate mastery of the computer and statistical packages as well as an understanding of the results of running programs on real data. This section of the portfolio has additional value as a resource when students must analyze numerical data in future courses and on their thesis or dissertation.

The projects engage students in authentic, real-world applications of statistics. The corrected exams, on the other hand, measure the same knowledge in a more isolated, objective manner. The reflective journal is a measure of how well students have learned to analyze their strengths and weaknesses. Do they understand how they learn best and have they attempted to utilize those methods more while applying skills to projects and exams? They are encouraged to write in the journal weekly. At a minimum, they are expected to write entries on their learning experience on the other sections of the portfolio. Some of the sections take just one or two weeks to complete, some take all semester, and some are group projects which are completed outside of class. The following sections represent the learning goals for the course.

    I   Mini Research Paper
    II   Computer Programs and Interpretation of Output
    III Group Project
    IV Midterm Exams
    V   Reflective Journal

Before: Preparing students for the use of portfolios.

At the beginning of the class, the portfolio is explained as a part of the course requirements. Many students are not familiar with portfolio assessment and even more have never been required to reflect on their learning and correct papers and exams after they have been graded. A lengthy discussion sets the expectation and purpose for these activities. I have found over years of using portfolio assessment that graduate students who are also fully employed professionals perform better with clearly delineated expectations. Therefore, the portfolio sections and criteria are established at the outset rather than allowing students to choose entries to the portfolio. A handout is distributed which outlines the contents and establishes the scoring rubric. The rubric gives the students a narrative description of the scoring criteria which communicates what they will have to do to receive the maximum, medium, and minimum number of points on the portfolio. Exemplars are distributed to reinforce the criteria and use of the rubric (see Lajoie, Chapter 14).

During: Issues in implementing a portfolio system.

Often during the semester, students are given an opportunity to discuss and ask questions regarding the portfolio. Students who have not been journaling hear from their peers how this activity has helped them. Examples of a more productive use of study time and learning resulting from correcting projects and exams are motivating to students who have not been as successful in these activities. Students are asked to share with others in groups what they have in their portfolios to date.

Students are also encouraged to turn in their portfolios during the semester for feedback from the instructor. This informs both the teaching and learning process and helps students to understand that they are responsible for constructing their own meaning out of the course material. Writing encouraging comments as well as criticisms is important in order to motivate and keep the students actively involved in the learning process. The following sections have evolved over time and seem to work well in this graduate level statistics course.

Section I: Mini Research Paper. One of the first learning tasks I assign in my graduate statistics course is in the area of nonparametric statistics. As already mentioned, learning and assessment tasks should be parallel in content and context and in some cases a learning task may be extended into an assessment task. This learning task evolves into an assessment task and, as part of the portfolio, becomes a baseline of comparison for later work as well as providing the first opportunity for reflection in the journal on the concept of statistical significance.

The task involves the student in choosing a problem or question which can be answered by collecting binomial data. The data must represent either yes/no, present/absent, success/failure or any other response which has only two possibilities such as heads and tails on a coin. Students formulate the question, collect a quick sample of data which should be obtainable in one to two hours, and then write an abbreviated research paper. I hand out several examples of the paper which includes an introduction of the problem, a null hypothesis, the methods used to collect the data, the findings section which consists of a visual representation of the data, and the conclusion.

The resulting paper as described above is presented in class and then handed in for initial feedback. The oral presentations can be done as a whole group in a small class or in groups of five to six in a larger class. I ask students to write a conclusion to their research although they are not trained to test for significance of the findings and the class discussion is guided into questioning the validity of their conclusions. Then I introduce a simple nonparametric statistical test, the binomial or sign test, and a six-step hypothesis testing framework. We work several sample problems together in class using student data. They learn how to make the determination of a significant difference between the two data points. The assessment task which goes into the portfolio is the research paper which they have edited and revised. This includes conducting the binomial test on their data and presenting the results in the findings section, writing a new conclusion, and editing the initial paper reflecting the comments I have made.

I begin the class with the mini research project to dispel the fear many doctoral students have that the intellectual demands of doing research are beyond them and most of all that they will not be able to understand and perform the statistical procedures necessary to complete the coursework. Through the mini research paper they discover the research process, the parts which logically present research to others, and the concept of significance.

Section II: Computer Programs and Interpretation of Output. During the semester students learn how to operate in the mainframe environment and run tests on two statistical analysis software packages, Statistical Analysis System (SAS) and Statistical Package for the Social Sciences (SPSSX). The second section in the portfolio contains SAS and SPSSX programs and listings. They learn to use SAS to run both descriptive and inferential statistics. Some of the programs they learn are procedures for means, correlation, analysis of variance, general linear models, and regression analysis. The number of programs taught on SPSSX is fewer. I find this package particularly good in describing data and presenting it in a way which can be transferred into a written document. Some of the procedures they learn on SPSSX are frequencies, tables, and reports.

The learning process is guided by several steps which individual students begin to omit as they become comfortable with the procedures. First they are introduced to the program in class both off and on the computer system. I hand out the printed program and explain each line in the program and what it tells the computer. We then calculate the most simple statistics by hand so that they know what the results of the program will look like. We then move to the computers and they learn to get on line on the mainframe and copy the prepared program off of my instructor disk onto their student disk and run the program. They then learn how to access the listing, the results of running the program, and print out both the program and the listing.

Each program is then practiced using different data sets which the students must enter into the program and run successfully. One of the programs which they have written and the resulting listing is put into the portfolio as an example of each procedure. When the course is complete they will have produced six to eight SAS and three SPSSX programs and listings.

Next I give them a research problem and some data and ask them to work in groups to decide what the hypothesis is and what analysis programs they need to run. They then work on the computers in groups to write the new programs, run them, interpret the listings and present the results in writing. This is practice for the group project.

Section III: Group Project. In the group project the students choose which type of design and statistical test they want to use. Each group writes a hypothetical problem, either collects or constructs the data, and chooses the type of analysis which is appropriate. They must make several decisions regarding the data set, what data to "collect," how to enter and label the data, and how to analyze the data. The entire study is then written up by the group and the results presented to the class. This group project goes into the portfolio.

Section IV: Midterm Exams. The midterm exams, of which there are two, are also placed in the portfolio. After the exam, the students are to analyze their own performance, correct wrong answers, and choose one area in which they need additional study. After the students have had a week or so to correct the exam, I review each response and answer any questions which remain. They enter into the portfolio some evidence of their study and the corrected responses to the questions on the exam which presented difficulty. In some cases, students may need to study more than one area of the exam in order to make the corrections. If they provide evidence of extensive study they earn bonus points on the portfolio which in turn may improve their grade.

Section V: Reflective Journal. One of the most important aspects of the portfolio is the reflective journal. As each assessment task is entered into the appropriate section of the portfolio, I ask them to analyze their work. Some of the questions I ask them to answer are 1) What did you have trouble with in producing this product? 2) How did you change the product after the initial feedback? 3) How could the product still be improved? and 4) What was the most important thing you learned in producing this product?

Additional questions for self-reflection on the group project include: Did you work with someone? How equitably did you contribute to the joint project? Did you help someone else? Who did the work on the computer? What problems did you encounter? How did you solve them? Did someone give you help?

The self-reflection required by this section involves recognizing the processes of metacognition which are stimulated. Bringing mental cognition out into the open through written reflection helps to teach students the mental processes they use effectively. Some of the thinking processes they should exhibit and recognize include analysis, synthesis, and evaluation and the application of these to the problem. Students' reflections can reveal much about the way they learn. By reviewing students' comments, the instructor may gain insights into the most effective and least effective teaching strategies they use. Here are a few examples which typify entries in student journals.

> At this point, had a glimmer the study was not addressing means so went back to my handouts and re-read chi square. Next tried to use a goodness of fit approach and realized attributes were not dichotomous so my search continued. As I was flipping through notes, AHA, remember correlation diagram and decided this was a correlation problem.

> Explaining <u>out loud **why**</u> ... listening to myself explain a topic or premise is a beneficial learning tool - no matter how time consuming. Teaching to the invisible student helps me clarify my knowledge or gaps of knowledge.

> Correcting the tests is a good learning process. A good review of what missed and makes you go back and review those areas.

> Use of a summary sheet acts like a self-quiz. By writing down what I think to be accurate in a visual table or format I can conceptualize better, check for accuracy easier, and reinforce progress to date!

> Self reflection and correction of exam is an excellent learning tool. Rather than dismiss my exam as over I developed a deeper understanding of material. I will continue this approach after this class and have recommended it to others.

> At this point, had a glimmer the study was not addressing means so went back to my handouts Six-step hypothesis testing is a sequential process which keeps me on track and helps to define significance and draw conclusions... I have rewritten examples from class to practice

this method. Another advantage to the 6-step approach is that it reduces my math anxiety.

Student comments emphasize the value of the reflective journaling. They tell their fellow students the value of this technique and continue to communicate its worth in office visits after the class is over. My comments on the journal focus on extending the students' thinking. This section of the portfolio is the most difficult to grade but probably of the most value to the student.

After: Interpreting and using portfolio information.

The students are encouraged to hand in the portfolio as soon as it is complete so that the grade and feedback may reach them before the final exam. They are also encouraged to make an appointment and discuss their learning process and progress on the material following the portfolio grading. The portfolio is awarded 300 total points which is about one third of the total for the course. In some cases, such as the first project and two midterm exams, the work received a grade when it was handed in originally. The revised version or corrected exam and the accompanying reflective journal entry is scored as part of the portfolio.

The points awarded in each section correspond to the scoring rubric. As described earlier, the rubric includes a description of the dimensions for scoring, the scale or points possible on each and the characteristics of the dimension expected at each level. Figure 1 (at the end of this chapter) provides an example of the general rubric used in scoring much of the portfolio. The dimensions in the rubric are statistical knowledge, strategic knowledge, and communication. The scoring rubric developed by QUASAR (Quantitative Understanding: Amplifying Student Achievement and Reasoning), a national project funded by the Ford Foundation, served as a guide for the development of the general rubric (see Lane, 1993).

Summary

The overall purpose of the sketched portfolio is for the student to demonstrate competence as a researcher. The completed product illustrates what the student understands and is able to do in the area of statistics. It also indicates where more study or reflection is necessary in order to demonstrate competence. The reflective journal is a key component of the portfolio due to the conceptual nature of statistics. Often a clear understanding of the concept only requires more processing time on the part of the student and the journal requires active mental processing.

The elements described in the portfolio sketch are examples of what might be included. The elements should match the desired course outcomes or critical learnings. Some other options include asking students to analyze the methodology of a study published as a journal article, critique claims in a newspaper article using survey or other statistically based information, or create their own article for a student newsletter or write a conference paper. The portfolio can be extended and adapted to any educational setting or student population.

IMPLICATIONS, CONCERNS AND BENEFITS

The complex practices and processes involved in portfolio assessment require a great deal of time and effort. Teachers must commit not only to creating learning activities, assessment tasks, and opportunities for reflection but must also be committed to using the results, being affected by them and by the insights into their teaching which result. There is also time involved in grading the portfolio. If you don't have time to read the student's work then don't choose portfolio assessment. If you have an assistant or grader, it is best to have the person grade the exam and grade the portfolios yourself. Bias is a difficulty in grading this type of assessment. If the class is small, you will know each person's work and their progress throughout the course. This means, as an instructor, you must be aware of any bias you have concerning individuals and take precautions that your bias does not enter into the grading process. One means of guarding against bias is by grading each portfolio and then going back through each one a second time to check for consistency in the awarding of points.

When students' knowledge results from an active transaction with the information being learned, it is best if portfolio entries are made throughout the year. In this way, assessments become embedded in the topics of study and reinforce hands-on teaching and learning (Shavelson & Baxter, 1992). This gives teachers and students opportunities for reflection and knowledge of what needs to be reinforced or reviewed throughout the course. The multi-source nature of portfolio assessment affords students more opportunities to demonstrate their knowledge and to show improvement. As teachers, we must model the complex habits of mind and ways of processing information which we ask of our students and ask for feedback from students on new activities which evolve from the process. This self-analysis has the potential to improve the strategies used in the classroom as well as the tasks. It gives the teacher more formative information on the effect of instruction on learning. As tasks are completed, the teacher can analyze strengths and weaknesses in teaching and make changes which will increase learning.

Another benefit of portfolio development is the dialogue which takes place between the students and the teacher. This dialogue begins with a discussion of what will be included in the portfolio as evidence of concept mastery. It may be an oral dialogue or a written one such as in student journal writing and teacher response. When students begin to write reflectively, there always seems to be discussion in class of what and how to write. The teacher needs to reinforce the part this process plays in students taking responsibility for their own learning. Students who by this practice continue to expand their understanding and construction of a cognitive web of statistical knowledge benefit greatly. The addition of the reflective process in the form of journaling facilitates this construction of knowledge. Knowing when to step in and redirect a student's efforts and when to step back and allow for the process of discovery takes experience with this type of discourse (Zessoules & Gardner, 1991).

Portfolio use can extend learning into the assessment process. Portfolio assessment is also extremely flexible and may be adapted for a wide variety of settings including K-12, post-secondary education, and graduate programs. The issues and concerns are different depending on the setting and use. In K-12 systems, there are many schools and school districts who have adopted portfolio assessment in various subject areas (see Brandt, 1992). Most experiences have been positive and participants have found that assessment projects engage both teachers and students in reflection on the learning process. In some cases, portfolios have motivated students and given their writing purpose. One teacher concluded that long-term situational math problems included in the portfolio were of the most lasting importance to students (Knight, 1992). The Portfolio Assessment Clearinghouse publishes the Portfolio News quarterly which includes descriptions of portfolio projects, articles on how and why to use portfolios, and reviews of the literature (Cooper & Davies, 1990 to present).

Portfolios have been developed at the state level. A 1990 survey by the Center for Research on Evaluation, Standards, and Student Testing at UCLA found that nearly half of state testing programs either had alternative assessments in place, were planning to implement them, or were actively exploring the concept. Alaska has encouraged innovative assessment projects and has assisted a number of districts in the area of portfolios. Several states are using portfolios for documenting work in mathematics and writing. Vermont and Kentucky have had 4th and 8th grade writing and mathematics portfolios since 1992. California initiated the use of portfolios as part of their performance assessment project in the 1993-1994 school year. These large-scale state efforts have encountered some resistance both from factions of the community against performance assessment and from teachers.

Some lessons have emerged from the large-scale portfolio assessment programs in Vermont and California. In a report of the pilot year of Vermont's Mathematics Portfolio Assessment Program (Vermont Department of Education, 1991), several suggestions for improvement of the system were made. Among those listed is the need for additional professional development and materials to be committed to teacher training. A recent Rand study found that, as a result of the training provided since 1991, teachers are teaching more problem solving but that there is still a need for more training (S. Rigney, personal communication, March 8, 1995). Due to some additional training on the structure and contents of the portfolio, the number of scorable portfolios has gone up considerably from the 83% of 4th grade and 58% of 8th grade portfolios reported in the earlier state publication. The inter-rater reliability of scoring remains an issue although it has improved over the four years of the program. It is now reported to measure $r=.80$ in mathematics and an $r=.70$ in writing. The Vermont program is continuing and the teachers and other participants seem to be committed. There is an understanding that it involves systemic change and as such takes time to implement successfully (S. Rigney, personal communication, March 8, 1995). The California project had only one sampling and scoring session prior to the discontinuation of the program in December, 1994. The issues which emerged from studying the results were 1) a question of whose work was represented by the portfolio, 2) the consistency of the pieces included, and 3) the reliability of the scoring (B. Thomas, personal communication, March 14, 1995).

The "whose work is it" question is of concern with all assessments completed over time. This question is not unique to assessment; any assignment done outside the classroom is open to the same criticism. There is also a controversy surrounding group tasks. If we believe in the merit and benefits of group efforts in learning, then this issue of "who's work" should not eliminate the inclusion of group assessment tasks in the portfolio. If the critical elements of cooperative learning are included in the approach to using groups, then the individual would still be held accountable for learning (Johnson, Johnson, & Smith, 1991). The important point is for raters to know the conditions under which the work was done. If you as the teacher act as the rater, then you would have this context information and the ratings would more likely be reliable.

Interest in alternative assessment is growing although the form the new assessment will take and many issues in the development are still being debated. Each instrument needs to be thoughtfully crafted and requires a specially developed or adapted scoring method. Portfolio assessment is just one form of assessment based on demonstrating outcomes. Because it is multi-source, some evidence of reliability is evident. The inclusion of both traditional and authentic sources of data complete the picture of the learner and help to direct learning to meet student needs. The portfolio guides the student through powerful, on-going assessment and provides a record of progress and a resource for future learning.

General Rubric		
Dimensions	Score Level and Points Awarded	Characteristics of Response
Statistical Knowledge	Level 3 Excellent Total points awarded project may be 50 to 60	Shows understanding of the problem's statistical concepts; uses appropriate terminology and notations; executes computations completely and correctly.
	Level 2 Acceptable Total points awarded project may be 39 to 49	Shows understanding of some of the problem's statistical concepts; uses nearly correct terminology and notations; may contain computational errors.
	Level 1 Poor Total points awarded project may be 0 to 38	Shows very limited or no understanding of the problem's statistical concepts; may misuse or fail to use terminology; makes major computational errors.
Strategic Knowledge	Level 3 Excellent Total points awarded project may be 50 to 60	Identifies all the important elements of the problem and shows understanding of the relationships between them; gives an appropriate and systematic strategy for solving the problem; gives clear evidence of a solution process which is correct.
	Level 2 Acceptable Total points awarded project may be 39 to 49	Identifies some of the important elements of the problem and shows some understanding of the relationships between them; gives evidence of a solution process which may be incomplete.
	Level 1 Poor Total points awarded project may be 0 to 38	Fails to identify important elements or places too much emphasis on unimportant elements; may reflect an inappropriate strategy for solving the problem; solution process may be missing, difficult to identify or completely unsystematic.
Communication	Level 3 Excellent Total points awarded project may be 0 to 38	Gives a complete response with a clear, concise explanation or description; presents supporting arguments which are logical; may include an appropriate diagram and examples.
	Level 2 Acceptable Total points awarded project may be 39 to 49	Gives an explanation or description which may be somewhat ambiguous or unclear; may include a diagram which is flawed; arguments may be incomplete or may be based on an unsound premise.
	Level 1 Poor Total points awarded may be 0 to 38	May have some satisfactory elements but may fail to complete or may omit significant parts of the problem in the explanation or description; may include a diagram which incorrectly represents the problem, is difficult to interpret, or no diagram.

Figure 1. Portfolio Scoring Rubric

Chapter 14

Technologies for Assessing and Extending Statistical Learning

Susanne P. Lajoie

Purpose

This chapter does not provide an extensive overview of the use of technology in statistics. Rather, a detailed summary is provided of how one researcher has used technology to teach and assess statistics in grade eight. Both traditional modes of assessment as well as technology-driven methods will be described in an attempt to demonstrate how multiple mediums of assessment can be used to provide a profile of students' statistical knowledge. The research reported here is grounded in cognitive theory with an emphasis on theories of learning that emphasize learning situations that are concrete rather than abstract. In other words, students learn better by "doing" statistics rather than just computing or reciting statistical equations or definitions. This research program is designed to provide authentic learning and assessment situations (Lajoie, 1995). The term authentic refers to meaningful, realistic tasks and assessments that validly assess what the learner understands.

INTRODUCTION

One of the lessons of the reform movement in mathematics has been that instruction and assessment can be thought of as a dynamic (as opposed to static) process whereby new forms of instruction lead to new forms of assessment, and that the results of student assessment can inform teachers in making instructional decisions to improve student learning (National Council for Teachers of Mathematics (NCTM), 1989; 1995). This chapter will provide a profile of what assessment methods can be used to monitor student progress in statistical comprehension and skill. In this regard, examples with the use of technology are provided so that teachers can see ways in which assessment methods can be developed given their resources.

There are both benefits and obstacles to the use of technological resources for teaching and assessing statistics. The author is not suggesting that technology is a preferential form of instruction in this domain but rather, that it is a useful alternative. Computer software, computer simulations, multi-media technologies, computer-based learning environments, and the internet are becoming more affordable and are being used for educational purposes. The real question is whether or not technology can improve statistics education. This question is addressed by reviewing how technology can increase understanding and facilitate assessment. Obstacles to incorporating technology are also reviewed.

Technology and learning with understanding

The essence of technology, as used in this paper, is that it is a tool or set of tools that can facilitate teaching, learning and assessment when designed to do so. Technology can be an ambiguous term when applied to education since it can refer to such things as the use of chalk, paper, pencil, TV, video, calculator, and computers (Hawkins, 1996). Technology, as used in this paper, refers to the use of computer software that is either commercially produced to assist students in graphing or analyzing data, or software that has been designed to make instructional and assessment goals more visible to students through multi-media technologies, or software that can be used to record students while they use computers to solve statistics problems. Technology, as described here, goes beyond simple game-like drill and practice activities and instead is used to teach understanding when pedagogical principles guide its use in classrooms (Perkins, Crismond, Simmons, & Unger, 1995).

Understanding emerges from mental activities that involve constructing relationships, extending and applying knowledge, reflection, articulation, and making knowledge one's own (Carpenter & Lehrer, in preparation). Technology can assist students' understanding of statistics by serving as a partner to students while they engage in such mental activities, helping them accomplish things they might not be capable of doing without such assistance (Lajoie, in preparation; Lajoie & Derry, 1993; Salomon, Perkins, & Globerson, 1991). Every cognitive task consists of many processes that compete for attentional resources from a learner. Computers can share some of the attentional burden by performing lower order skills, such as drawing a graph, while students' resources are free to think about the meaning of the graph. Cognitive overload can be avoided if the technology is user-friendly (easy to use) as opposed to user-unfriendly (difficult to use whereby a learner spends more time learning how to use the computer than how to reason with statistical data).

One of the most basic ways that technology can be used to facilitate understanding is by acting as an intellectual mirror for students to peer into while problem solving (Schwartz, 1989). The more interactive the computer technology the more it can help students reflect on their problem solving processes. Within the context of statistical investigations students can get instant feedback on their hypotheses when they enter their data into the computer for graphing or analysis. Positive feedback would result in the graphs or analyses of interest, whereas negative feedback would results in graphs and analyses that were not expected. Technology can also facilitate reflection though "procedure capturing" (Kaput, 1992) whereby student computer actions in the context of solving problems could be replayed at the completion of problem solving. Such "replays" help learners focus on the processes they used to solve problems. Computers can capture an entire sequence of student actions, some of which are plans, and some of which are actions, and replay these actions for the student to reflect on (Lajoie & Lesgold, 1992). Some computer-based learning environments replay student actions and let students compare their actions with expert solutions so that better models of performance are available for inspection. These same types of environments can be designed to be adaptive to individual differences by providing assistance or coaching to those learners who need help. Procedure capturing, tracing, and replays can be an effective instructional tool since externalizing thought processes and actions helps students monitor their own performance. In essence, this monitoring is a form of self-assessment where students learn to assess themselves and become more independent. Teachers benefit from such tools in that they can review student traces to assess student difficulties and make important instructional changes that can remedy these difficulties.

The dynamic nature of technologies makes instruction an active and often student-driven activity where multiple examples are created dynamically by the learner's own actions rather than instruction that is static whereby pre-established sets of examples are selected by the teacher (Kaput, 1992). Kaput suggests that technologies provide a mechanism for dynamically externalizing mathematical notations. He describes many types of external notations used in mathematics (algebraic, graphs, tables) and discusses ways in which technologies can be used to externalize mathematical notations in a dynamic manner so that students can focus on how their own actions affect changes in such representations. Kaput (1992) describes the impact that dynamic vs. static notations can have when students are learning about statistical variance. In static mediums, variance can only be described through multiple examples, whereas in dynamic situations, students can interact with the data and see the immediate impact that changing one data point could have on, say, a histogram or scatterplot (Hancock, Kaput, & Goldsmith, 1992; Konold, Pollatsek, Well, Lohmeier, & Lipson, 1993; Rosebery, & Rubin, 1989). Kaput (1992) suggests that manipulating objects or mathematical entities on the computer can be more effective than physical manipulatives when the computer makes the links between such notations and mathematical models. Simply observing something happen on a screen will not guarantee that the learner has made the connection between, say, a graphical representation and the meaning behind the representation. It is often necessary to make the direct connection for the learner. Still it seems apparent that technology can empower students in the use of statistics by letting them actively engage in "doing" statistics (Hancock, et al., 1992; Konold, et al., 1993; Lajoie, Jacobs, & Lavigne, 1995; Lehrer & Romberg, 1996; Rosebery & Rubin, 1989; Scheaffer, 1988). Technology can be adaptive to individual differences in learning specific content by providing multiple types of representations; i.e., hypertext, graphics, animations, and videoclips and can remedy student misconceptions by providing alternative views of a phenomenon or through computer simulations (Nickerson, 1995).

Technology and assessment

From an instructional perspective, teachers sometimes struggle to find ways to clearly demonstrate what their expectations are of students. It is often difficult to demonstrate abstract concepts and consequently to tell students how they will be assessed in such contexts. Teachers' expectations can be made clear when technology is used to post concrete examples of good performance. When such examples are made available to students, students can better attempt to meet the teachers' expectations, thus facilitating both instructional and assessment goals. As new assessment techniques are being developed for statistics, it is necessary to consider how technology can help teachers manage the assessment process. Just as students will need time to adapt to new technologies in the context of instruction, teachers will need assistance in learning how such technologies can be used to facilitate instruction and assessment.

Obstacles to practice

Although there are benefits of technologies, there are also obstacles to incorporating them in the classroom. Kaput (1992) reviewed many of these obstacles in mathematics classrooms; i.e., lack of physical resources, such as phone lines for internet hook-ups, or up-to-date hardware and software, and most importantly lack of teacher training in technological resources. The pedagogical obstacles are the most severe. Teachers often do not have the necessary training in teaching statistics or in the use of technology in this context. As Hawkins (1996) points out, there are too many choices that teachers must make with little guidance; i.e., which are the more appropriate statistical and graphical packages. She suggests that although these new technologies provide speed and access to larger data sets and more complete access to the entire statistical investigation process, attention must be paid to subtle problems that can occur with these tools. For instance, automatic scaling procedures may reduce the learners' awareness of how scaling affects the graphical representation and subsequent interpretation of data. Teachers must draw students' attention to these features. Further, there are times when the speed of access that technologies provide may lead to what Ben-Zvi and Friedlander (1996) referred to as uncritical statistical thinking, where students generate a large number of statistical analyses and graphs but their focus is on extrinsic or aesthetic features of these activities rather than the underlying statistical meaning. On the other hand, Ben-Zvi and Friedlander (1996) found that these same technologies could lead to the meaningful handling of data and metacognitive skills related to statistical reasoning.

In summary, technology can be used to improve statistical instruction and facilitate student understanding and assessment by providing active interactive learning opportunities where multiple representations, and dynamic statistical notations, can be used to extend one's comprehension of statistical data, graphs and analysis, and by confronting misconceptions head on with appropriate simulations that shed light on statistical concepts. Technology can also externalize the learners' problem solving processes in a manner that could help individuals monitor or assess their own performance. As an instructional aid, teachers benefit from technology, since some instruction can be dynamically created through student interactions with technology. For example, student input can result in multiple examples that are constructed dynamically based on their input rather than instruction that is pre-established based on a contained set of statistic examples. Teachers may have more time to interact with individual groups of students and observe their statistical investigations. Through these observations and by reviewing student problem solving traces, collected and stored by the computer, teachers can change their instructional strategies to fit the needs of their students. The Authentic Statistics Project (ASP) described below, uses technology for both instructional and assessment purposes.

THE AUTHENTIC STATISTICS PROJECT

This chapter describes a technology implementation conducted as part of a research project rather than as part of a pre-existing statistics curriculum. A pilot of this project was conducted based on a partnership that was formed with a mathematics teacher who taught two grade eight classes. The teacher reviewed his curriculum content and placed our statistics instruction following a unit on graphing. In so doing he was able to build on students' prior knowledge of graphing by relating it to data presentation. Once this link was made, students were introduced to computer tools that would assist them in constructing graphs and performing analyses in the context of their statistical investigations. The use of technology alone will not ensure the development of statistical reasoning. Just as in other instructional settings a theory of the learning and instructional process must guide the use of technology. The pedagogical model that guided the ASP was the cognitive apprenticeship model which includes six methods for developing an optimal learning environment: modeling, coaching, fading, articulation, reflection, and exploration (Collins, Brown, & Newman, 1989). What follows is a description of how this model guided the design of ASP, and how the use of technology facilitated this approach to statistical instruction and assessment. This description will be followed by a detailed look at how the ASP project could be changed to better make use of the strengths of technology, and to better help teachers in the classroom.

Modeling, coaching, and fading

Modeling, coaching, and fading represent an interconnected view of instruction and assessment. Instruction may consist of modeling the knowledge and skills you want students to learn. However, in order to coach or assist learners in the context of such models, one must assess the learner's level of skill acquistion and provide the right level of assistance for that learner. Fading of such assistance occurs when teachers' assess that these same learners no longer need assistance.

ASP was conducted over a two-three week period during the regular mathematics class. Eight Macintosh computers were brought into the classroom for the duration of the project and three to four students worked at each computer workstation. Each group had access to either the teacher, or graduate student mentors, who were available to facilitate the statistical investigation process using computers. For the sake of simplicity, teachers and mentors will be referred to herein as teachers. Each teacher was responsible for modeling the skills and knowledge needed to carry out the statistical investigation processes. Modeling refers to demonstrating skills or knowledge. There were two phases to our project, the knowledge acquisition and the producer phases, and hence depending on the sequence of instruction different things were modeled. During the knowledge acquisition phase students were taught basic statistics concepts hand-in-hand with procedures for "doing" statistics. Students were introduced to statistics through a tutorial that taught them factual knowledge in the context of solving statistical problems with computer tools. Since this was a pilot study and we were still trying to identify what aspects of ASP would be effective, the tutorial was in paper form. Each group was led through a standardized set of instructions that would help students understand statistics as well as how to use the statistical package MyStat™ and a graphical software package Cricket Graph™ to solve statistics problems. The teachers were available to coach students whenever difficulties occurred while carrying out these tasks. The factual portion of the tutorial introduced students to a new vocabulary of statistical concepts that dealt with a basic understanding of measures of central tendency. Statistical concepts and graphic representations were modeled by first providing concrete examples of such things from local newspapers. For example, histograms, pie charts, percentages were demonstrated in the form of voter polls, sports statistics, and box office grosses for popular movies. Next, the types of procedures, or tools, that were available on the computer, i.e., the MyStat and CricketGraph applications, were modeled. These applications provide opportunities for representing the data in a multitude of ways, statistically, numerically, and visually.

Mini-experiments served as instructional activities to teach the students concepts of central tendency and variability. These mini-experiments serve to situate and extend the learning through what Bransford, Hasselbring, Barron, Kulewicz, Littlefield, & Goin (1989) referred to as a macrocontext; complex situations that require students to formulate and solve a set of interconnected subproblems. It was through these activities that the various software tools were modeled. The first mini-experiment included data gathering activities on Pulse Rates. Each group collected pulse rates in their group and entered their data into the computer and analyzed it using MyStat. Individual group data was then compared to data from the whole class. Discussions about the mean, mode, and median, data, sample, randomization, population, and range were generated in the context of these experiments. Following this exercise another group of students "ran-on-the-spot" for a minute and students collected, analyzed and interpreted their data by comparing it with the "at-rest" data. By comparing two group means predictions were made, hypotheses were tested, and results were interpreted. These experiments required whole class participation. Students were coached by instructions in the tutorial and by the teachers in how to use the computer to enter, analyze, and graph data. After completing the tutorial (knowledge acquisition phase), the producer phase began where each group was asked to construct its own statistics project.

The projects included several components of statistical problem solving: designing a research question, collecting data to answer the question, representing the variables and data graphically, analyzing data through statistical methods, interpreting data, and communicating understanding through class presentations. Since students worked in small groups to design their projects, much of the modeling and coaching would take place within the group when students shared different perspectives on the learning task. The type of modeling and coaching by teachers changed at this point in time. Teachers faded the type of procedural help (i.e., how to use the computer software) since students were adept at this after the tutorial, and increased the amount of modeling and coaching regarding such things as data organization needed for entering data into the computer in a manner that was conducive to the research question at hand. However, the most valuable type of modeling that occurred in the production phase was provided through the use of technology, in our library of exemplars.

Library of exemplars as a model for assessment

A computerized library of exemplars was developed (using HyperCard and Quicktime) as a means to demonstrate teacher expectations regarding how students might perform when creating a statistics project. The library of exemplars served to model these expectations by making assessment criteria visible to learners prior to their engagement in the statistical design phase. Examples were presented of both average and above average performance on each statistical component that students would be assessed on when they developed their own statistics projects (Lavigne, 1994; Lavigne & Lajoie, 1995). Technology was used to make assessment criteria more concrete in an effort to make it is easier for students to achieve these goals. Several components of the statistical investigation were modeled: what a research question looked like, how to collect data to answer a question, how to represent variables and data graphically, how to analyze data through statistical methods, how to interpret such data, and how to communicate about such projects in the context of class presentations. Each component in the library had concrete examples of what such criteria mean and how they are assessed. Although technology is the medium that is used for modeling the performance criteria, real students are actually modeling the performance criteria .

The exemplars were constructed by selecting video-tapes of students who had participated in a prior study and demonstrated different levels of performance on the above statistical components. Weaker and stronger examples of student performance were selected for each component so that new learners could compare these examples as they pertained to assessment criteria. Figure 1 illustrates the data analysis component. A textual description of the data analysis criterion appears on the left side of the screen, explaining what data analysis means in the context of a statistical investigation, and the maximum number of assessment points that students can get for this criterion. For each statistical component two video examples are available on the computer. To select the average or above average performance example students click on the respective video camera icon to see a video clip of how other students performed on that component. The average example illustrates a group that lists the minimum and maximum scores and the range. The above average example focused on the interpretation of their data. For instance, when asked whether they thought they would get the same results if they conducted their survey in Toronto as opposed to Montreal, the group responded that they were sure that different results would be obtained in Toronto, because they might market the products differently in another city. These examples help students engage in dialogues about what data analysis means in the context of previous student work and how such work was assessed. By making the criteria clear to students, all students could benefit by sharing a common understanding of the problem.

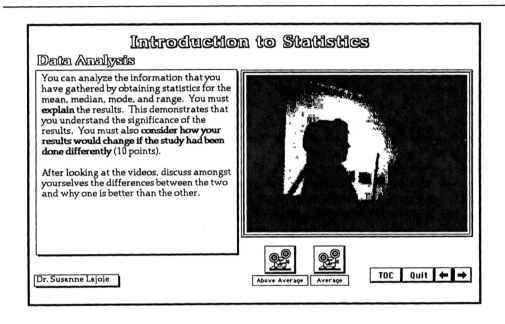

Figure 1. Data analysis exemplar

Reflection and self-assessment

The library of exemplars demonstrates how teachers can use technology to instruct students as well as to make their assessment criteria clear. Once students are given models of performance they can internalize the teacher's instructional goals and assessment criteria prior to constructing their own statistics projects. Furthermore, they can return to the computer at any time to refresh their memories of what a particular statistical concept means. Students can use such criteria to assess their own progress and compare their work with others. The use of technology in this case promotes students self-assessment and reflection on the cognitive components of the statistical investigation process. But how do we assess if students have internalized such criteria? One way that self-assessment was promoted was to ask each group to assess themselves on the same criteria they saw in the library of exemplars and to assess every other group's presentation on the same criteria. Assessment sheets were distributed to the groups with the maximum number of points available for each statistical component. Teachers independently assessed each group in order to monitor whether or not students' self-assessments were in accord with their own assessments of student work. After these assessment activities were completed, student self-assessment ratings were compared with teacher ratings, and ratings by other groups, and high correlations were found between the three types of assessments. These correlations tell us three things. First, students are not over-inflating their assessments in order to get high grades, rather their scores are in alignment with independent raters. Second, students are evaluating other students' performance fairly as opposed to under-rating other group's projects, which would indicate the negative effects of competition. Third, there is alignment between what is modeled and what is learned, or what is instructed and assessed. Technology, then, is useful in making teacher expectations clear and concrete enough for students to reflect on, internalize, and meet such expectations to the best of their ability. This exercise verifies the assumption that when teachers communicate their goals clearly learning will be facilitated.

Articulation and assessment

Students in ASP articulate their understanding of statistics in several ways. The most common form of articulation is a verbal one where individuals explain their reasoning with words, either orally or textually. In ASP, students verbally articulate their knowledge during the construction of their project where they must communicate their reasoning to their peers about a certain research hypothesis or data collection procedure. Students share their understanding amongst themselves and critique each other's ideas during the production phase. Groups also keep written journals documenting their statistical projects. Students also orally articulate their statistical comprehension during their presentation of their project, where they defend their reasoning about their hypothesis, their data collection procedures, choice of graphs, statistical analyses and interpretations of their findings. Articulation with technology is necessary in both these phases since the context of their statistical problem solving revolves around the computer as a tool used in their investigations. In fact, the liquid crystal display panel was used during student presentations to demonstrate different aspects of their statistical projects.

A less common form of articulation is what Kaput (1992) referred to as dynamic external notations. For example, the computer kept track of each group's use of the computer for graphing and analyzing their project data. All computer work was saved in the form of Screen Recordings (a Farallon product) which were exact computer-films of the sequence of activities that the students performed each day. An entire session on the computer could then be replayed like a videotape, only the video captures the procedures students used in the context of developing their statistical investigations. Teachers could view these recordings to examine whether or not these external representations were documenting students' progress in statistical reasoning. The dynamic external notations could be later used for assessment purposes to document changes in statistical reasoning.

The types of actions teachers considered important in terms of assessment of these external representations were: the number and types of graphic manipulations (i.e., pie charts), number and types of statistical manipulations (i.e., frequencies), the types of files that were accessed to develop projects, the creativity with which they personalized their presentations, and the types of difficulties students encountered in using the computer (i.e., attempts to run a statistics operation but not completing the operation for lack of knowledge). Off-task activities refer to activities such as playing with computer games or files that are not associated with the project.

Stiggins (1987) cautioned against the subjective interpretation of students' performance ratings and Linn, Baker, and Dunbar (1991) suggest that teachers be trained in a manner that would allow for some calibration in scoring performance data. We followed these suggestions by preparing a computerized demonstration of each of the types of actions students performed in the context of developing their statistics projects, as a way of assessing these external notations. This demonstration helped teachers be consistent in their assessments of such notations. Figure 2 describes the perform statistics criterion and what sorts of notations to expect. Text is used to describe the criterion and then teachers select PlayMT to get a screen recorded example of what performing statistics looks like in the context of ASP.

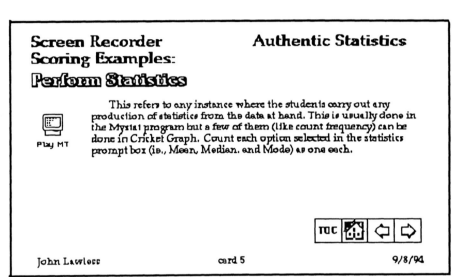

Figure 2. "Perform Statistics" criterion for scoring screen recordings

Along with the video clip are sound clips that describe the student activity and textual clips further elaborating the student actions. Together, the oral, textual and animated displays of each criterion help teachers score student notations in a more consistent manner.

The computerized demonstration serves to assist teachers in making decisions about what to assess and how to assess the students' problem solving capabilities. Once again, such assessment criteria for considering students' progress throughout their projects could be negotiated in that certain components may be more important than others, some added or deleted. The advantage of making the criteria visible through technology is that teachers can see contextualized video-screen segments of statistics work rather than single snapshots of performance. The screen recordings of students' progress in statistical investigations can tell us whether there is a relationship between types of notations students make and their actual understanding of statistics. Multiple forms of student data must be examined together in order to see the relationship between the type and number of graphs constructed and the final statistical presentation where justifications of student projects are communicated. In other words, we need to compare what students do with how they justify or reason about what they do in the context of statistical problem solving. Without such information it is impossible to say whether or not the different number of graphical representations students construct with the ease of technology demonstrates understanding or confusion.

Exploration

Students were given freedom to explore different aspects of the statistical investigation process and in so doing developed their own research question, collected their data, graphed, analyzed and interpreted such data. Computer tools were useful, in this regard, since students could generate and test their hypotheses as to what graph or what type of analysis might be appropriate for their question. The flexibility of the computer allows students to quickly find answers to the multiple hypotheses they may have at any particular time. One caveat, of course, is that students must understand how to pose such questions in a way that can be answered by the computer analysis. In other words, students must understand how to organize their data according to the type of analysis they choose to answer their questions. Teachers served as coaches, helping students when difficulties arose in the context of their statistical investigations. Given proper instructions, students can use computer software to graph their data in much less time then it would to draw a graph freehand. The speed of computer frees up time for students to reflect on the significance and interpretation of the data in the graph rather than on the construction of the graph. However, as mentioned above, student assessment must include evidence that students are indeed reflecting on the meaning of graphs and not just picking a number of graphs at random. Exploration as we see it should still be guided and monitored by teachers so that dead ends are not pursued and misconceptions are not reinforced.

SUMMARY AND IMPLICATIONS

This chapter provided an example of how technology promoted understanding of the statistical investigation process and improved assessment of statistical reasoning. The use of technology does not guarantee the coupling of instruction and assessment in statistical contexts; a theoretical framework should underlie the design of the use of technology in the classroom. The cognitive apprenticeship model is just one model that can be useful in this regard.

The six methods of the cognitive apprenticeship model were described to provide a framework for understanding the instruction and assessment process in the ASP project. Technology was not the sole instructional dispenser or assessment device in ASP. Rather, technology provided a context for facilitating student exploration of statistical problem solving through the pursuit of the multiple types of representations available to students for graphing or analyzing their data. Exploration in both the knowledge acquisition and producer phases of ASP was guided by teacher interventions or coaching when needed, and the fading of such assistance when no longer needed. Coaching and fading of assistance demonstrate the connection between instruction and assessment. Technology was successful at making the teachers' instructional expectations and assessment criteria clear through the library of exemplars. The computer was used to model the statistical components of problem solving by demonstrating how other students performed on such criteria. This library promoted students' reflection and critique of their or other student projects, and facilitated subsequent self assessments of their own projects. Furthermore, this use of technology helped align the teachers' assessment criteria with students' self-assessments criteria.

Technology, as used in ASP, was highly interactive, responding to students' requests by performing graphical and statistical analyses on the student data. All of the students' computer work and activities were recorded as screen recordings that could be played back and evaluated by teachers. The consistent or calibrated evaluation of this computer work was facilitated by the design of scoring templates. These provided teachers with multi-indexed examples of the types of activities and processes, including graphical and statistical manipulations, that could be assessed. There is much work to be done before technology-based projects such as ASP become fully implemented in the classroom. From a practical perspective, teachers need full access to technology in order to replicate or creatively add to the ASP work. Also, statistics must become a standard unit in the mathematics curriculum in order for teachers to allocate time and resources to this form of instruction.

Given our pilot data, improvements can be made to both the instruction and assessments of statistical problem solving. One improvement would be to use computers to standardize the knowledge acquisition phase so that the computer could take a key part in administering instructions and coaching, and in giving feedback. The computer could fade assistance when students have mastered the skills needed to manage their statistical investigations. Teachers can be most effective in coaching students in the production phase, helping students brain-storm about research questions and about how to collect, represent and analyze their data.

We know that technology can be used to effectively model components of the statistical problem solving process, and facilitate independent explorations and self-assessment. We need to spend more time thinking about the types of statistical knowledge that can best be exemplified using computers. For instance, Bright and Friel (in press) have examined how students develop graphical understanding. Results from such work are critical for designing more effective uses of technology in teaching and assessing statistical reasoning. If we understand how transitions in statistics learning occur, simulations can be designed along with direct instruction where key concepts and processes are modeled for the learner.

A single form or type of assessment is not adequate for representing statistical problem solving. Rather, multiple types of assessment must be combined. In ASP, oral forms of articulation were examined along with the recordings of students' computer actions during problem solving. Note that the computer can collect observations but it is the teacher who must interpret such evidence in order to understand and document student progress over time.

Perhaps as our understanding of how to assess these recordings or notations evolves, part of the assessment can be done dynamically by specialized computer programs. Methods must be found to ease the assessment burden of the teacher. Teachers need workshops that address both the cognitive aspects of statistical reasoning and how to assess changes in such reasoning, along with workshops that help them use technology in the classroom. At the same time, teachers must learn to use tools for managing complex forms of evidence of student learning and become adept at reliably interpreting such data.

Acknowledgments

I would like to acknowledge Nancy Lavigne for her assistance in all phases of this work, and Steve Munsie, and Tara Wilkie who have played a large role in this research. I acknowledge André Renaud who programmed the library of exemplars, and John Lawless who programmed the screen recording stack. I would like to thank Brent Blakely, Marlene Desjardins, Bryn Holmes, Cindy Finn, Steve Riesler, Jody Markow, Mike Thibault, Tina Newman, Litsa Papathanasopoulou for their assistance in various phases of data collection. I gratefully acknowledge the assistance of Phil Knox, a gifted mathematics teacher, who generously provided his classroom as a testing ground for this research. Special thanks to Thomas Romberg for his support of this research. Preparation of this document was made possible through funding from the Office of Educational Research and Improvement, National Center for Research in Mathematical Sciences Education (NCRMSE).

Chapter 15

Issues in Constructing Assessment Instruments for the Classroom

Flavia Jolliffe

Purpose

When constructing assessment instruments both the *purpose* of the assessment (feedback, grading) and the skills which are being assessed need to be considered. The main purpose of this chapter is to help teachers develop their own assessment instruments by giving specific examples of tasks. Unsatisfactory tasks are used to illustrate the pitfalls, and alternative versions are given as examples of good practice. Some comments on grading are included. The emphasis is on written assessment in the classroom, mainly of pupils aged about 14-19, but much is relevant also to introductory statistics courses at college and university level.

Consideration is given to ways of assessing factual knowledge, the ability to use computers, understanding of concepts and application of techniques, and communication skills. The pros and cons of multiple choice and open-ended questions are discussed as are the challenges of oral assessment and assessment of group work.

INTRODUCTION

In assessing student learning of statistics we need to assess factual knowledge, understanding of concepts, computational ability, appropriate application of techniques, and the practical skills of doing and communicating statistics. Although there are strong arguments for using practical work and projects as the principal methods of assessment of many of these qualities, other methods can be used successfully for most of them. Traditionally assessment has involved the student in answering written questions, possibly under a time constraint, but more recently some consideration has been given to the use of oral assessment.

Assessment can be formative or summative, a division corresponding approximately to the two main purposes of assessment, feedback both to the teacher and to the student (formative), and grading of the student (summative). In internal assessment in a classroom situation, which is the main concern of this chapter, feedback is more important than grading. All the points made, however, are also pertinent to external assessment by public examinations where grading is the only purpose.

Closed book timed examinations where students are required to do questions of a standard type have little to commend them as ways to assess student learning of statistics. Questions tend to be artificial in nature and are more likely to test short-term than long-term memory, and the time constraint may be unfair to some students. However, we can be fairly confident that the grading produced reflects the students' own performance. The provision of formula sheets reduces the reliance on memory and the extension to open-book examinations brings a new dimension into assessment and mimics the situation in the workplace. Some public examinations include analysis and discussion of real data (say one page of data) sometimes available to students before the examination. In other public examinations questions relate to newspaper articles (see also Chapter 10) or are based on reports of experimental investigations.

It is easier to see what is unsatisfactory with an assessment task than to start from scratch and write a good question, worded unambiguously in everyday language and meeting required objectives. Questions *can* be made crystal clear by using mathematical language of course, and we must use correct and precise terminology in teaching and assessment, but real life statistics questions are not posed in this way. Even when questions are subject to rigourous refereeing when being developed, students sometimes make a legitimate but different interpretation from that intended, or find unanticipated methods of solution. Bentz and Borovcnik (1988) consider explanations for alternative answers to relatively simple test items on probability. The issue must be more complex with more searching questions at a higher level.

This chapter starts the process of writing assessment tasks by giving examples of tasks and pointing out why some are poor and others are better. A general discussion of assessment issues is given in Chapter 10 of Hawkins et al. (1992). A framework for the generalisation of assessment tasks is given by Nitko and Lane (1991).

The design of assessment tasks should, as mentioned earlier, be guided by clear curricular objectives. For the purpose of this chapter, the following four goals were adapted from Hillyer (1979). These goals are compatible with and comprise part of the general framework of common goals presented in specific chapters in this volume.

1. To develop knowledge and an appreciation of the ideas involved when looking at information.

2. To develop knowledge and an understanding of methods of presenting information by charts and diagrams including those used in business and newspapers.

3. To develop necessary interpretive skills so as not to be misled by statistical information.

4. To develop a critical approach to the methods of collecting and presenting information and to the conclusions drawn from statistical data.

Below, I first discuss how to assess factual knowledge, then I discuss issues of assessment involving computers. After presenting aspects of assessing using multiple-choice and open-ended questions, I focus on the assessment of group work.

ASSESSMENT OF FACTUAL KNOWLEDGE

How can we assess factual knowledge? Why might we wish to do so? A fairly popular form of assessment question has been to ask for definitions of terms or formulae, particularly in class tests after pupils have been asked to learn material or in formal examinations; for example, "Define the median," "Give a formula for the arithmetic mean of a frequency distribution." Such questions test little more than the ability to retain material in memory and recall it correctly. There is not much point in this because in the workplace frequently used facts are learnt through their use and others are looked up when needed, but it *is* important to assess whether students know what to use and what to look up.

Let's start by considering testing factual knowledge of the median. We could ask for an interpretation; for example,

> The median number of hours spent doing homework by the 30 members of a class last week was five and a quarter. Write a sentence explaining what this tells us about hours spent on homework by class members.

Or we could ask when and why a median would be a suitable summary measure. Or we could ask a more specific question such as:

> In negotiating for a salary raise the union representative quotes the median income. The employer's representative quotes the average (arithmetic mean) income and this is higher than the median. Explain why these two summary measures can differ. Which measure do you think is more appropriate here, and why?

Rather than ask students to quote a formula it would be better to ask for a calculation, as they would need to know the meaning behind a formula in order to do this. For example, let's consider the arithmetic mean. The question below does not discriminate against those who have other ways of remembering how to find the mean than via an algebraic expression, such as those who use a mental picture of the layout of calculations. A formula sheet is not necessarily useful to such students.

> The following frequency distribution gives the number of days it took for a person to drive a certain distance. Estimate its mean to the nearest 0.1. Show any necessary work.
>
> | No. days | Frequency |
> |----------|-----------|
> | 0- 2 | 14 |
> | 3- 5 | 29 |
> | 6- 8 | 19 |
> | 9-12 | 10 |
>
> (Bernklau)

Another possibility is to ask a question which tests understanding of a formula or concept as in the examples below.

> Circle the *single* best answer.
>
> Leonard made an 84 in the first test and an 89 in the second test. If all tests are weighted the same and Leonard's teacher does not round averages, what does Leonard need to make on the third test to have an average of 90?
>   a) 90   b) 96   c) 97   d) 98   e) 100
>
> (Kaigh)

> A set of scores has a mean of 68.7. If 10 were added to each score then the mean would be ......
>
> (Bernklau)

There are good pedagogic reasons why students should have some familiarity with formulae and should do some simple hand computations. Both aid understanding and help remove the black box aspects of calculators and computers (Jolliffe 1991, 1993), but they should be seen to be short first steps and should not be laboured to such an extent that they become stumbling blocks. The principles behind methods of making computations easier, such as change of origin and scale in finding means and variances, are important but should be presented briefly.

Another way of testing factual knowledge is to ask for explanations of how to perform operations or the meaning of terms; for example, "Explain how to use a graphical technique to estimate a median from a grouped frequency distribution" or "Explain what is meant by simple random sampling." One problem here is that those who have memorised notes or portions of textbooks may be able to reproduce appropriate sections without understanding (and some might write out the wrong section!). Another is that giving explanations in the abstract is quite difficult. Thus a question asking for an explanation might be improved if it also asked for an illustration. For example:

> Explain how to use a graphical technique to estimate a median from a grouped frequency distribution. Illustrate your answer by estimating the median of the frequency distribution below which ....
>
> Explain what is meant by simple random sampling. Describe how you would select a simple random sample from the list of 200 people attending a one-day meeting.

An alternative to providing an illustrative example within the question is to ask students to describe examples with which they have been involved. This has the advantage of linking the assessment to practical statistics. But if students have access to the material they could just copy out examples; if they do not the question is mainly one of memory.

COMPUTATIONAL ABILITY

Computational ability is important and students need to practise questions on computation, but teachers should not set large numbers of such questions just to keep a class occupied, nor base a large proportion of assessment on such questions. Questions which do little more than ask for a computation— for example, "Calculate the standard deviation" or "Fit a regression line," —test whether students can choose the appropriate function, and do arithmetic or use technological aids correctly, but do not test ability to apply principles in statistics. Both factual knowledge and computational ability can and should be tested through questions which are of statistical interest.

Students should be shown how to obtain approximate answers with pen and paper so that they develop a feel for the magnitude of answers and can recognize when an answer is clearly wrong, perhaps because a wrong function button has been pressed on a calculator. Ways of testing whether students are able to do this include: asking for an approximate answer, presenting the question in multiple choice form, asking for an explanation as to why a given answer cannot be correct, and asking for eye-ball comparisons of data sets.

Students who get an impossible answer at any stage such as a negative variance, a probability outside the range [0,1], or a correlation coefficient outside the range [-1,1] or where the sign is clearly wrong, should be penalised severely if they do not comment that it must be wrong because this shows lack of factual knowledge and of understanding.

ASSESSMENT OF ABILITY TO USE COMPUTERS

As statistics packages (including spreadsheets) and word processor packages become more readily available in or out of the statistics classroom, students will be making increasing use of a computer for statistical work. Assessment of their ability to do so will therefore take on a greater importance. The type of assessment tasks need not differ fundamentally from those in which the use of the computer is not required, and all the principles of good and bad assessment apply here too. Many papers have been published describing experiences of using the computer in teaching and advocating its use, but few mention assessment. An exception is the paper by Phillips and Jones (1991) in which they state briefly how they aim to make assessment reflect the style and flavour of a statistics course for engineering students in which computing is an important component.

The skills we might wish to assess include knowing what analysis it is appropriate to do, the ability to get a package to perform this, and interpretation and editing of output. Assessment of knowledge of what statistical analysis is required is almost no different when computers are used as when analysis is done by other means, although the choice is clearly limited by what packages are available. This is likely to become less of a problem in the future as simpler packages become more sophisticated and computers become more powerful and user-friendly.

Assessment of practical skills such as importing data and other files, and being able to obtain a hard copy of output, possibly in edited form, might be assessed in computer science courses and study of methods used there could be helpful. Skills such as being able to access and use a package, including data entry, are more specific to statistics. Actual computing skills as used in spreadsheet manipulations or in writing routines to use within a package could possibly be assessed in the same way as a "hand" computation. Assessment of the selection of interesting and appropriate parts of the output and interpreting them should present no particular problems in that the role of the computer is that of a tool like a calculator.

The teacher needs to avoid the temptation to overwhelm students with data. Just because packages can deal easily with large data sets does not necessarily mean that it is good for students to spend a lot of time looking at large data sets. Students do not have to do all the computing themselves—they can be asked how they would use a package and to say what commands they would choose, and they can be asked to interpret a pre-prepared piece of output. The following is an adapted example of a question which included computer output showing commands, dotplots and stem and leaf diagrams.

Fuel consumption figures for 57 types of car are reported in miles per US gallon in the following Minitab output. The data come from issues of the US *Consumer Reports* magazine between 1987 and 1989. The data show fuel consumption for city driving, an open road trip and expressway driving on an interstate highway at a steady 55 miles per hour.

(a) From the data shown, what is the most economical fuel consumption achieved with expressway driving? What is the median fuel consumption for open road trips?

(b) Write a brief report (no more than 200 words) summarising the conclusions you draw from these data about fuel consumption, covering both the way it varies between different types of cars and the way it is affected by the type of driving done.

(Fuller)

Use of computers in an exploratory way, for example, building up sampling distributions or seeing the effect of outliers on regression lines, would not normally be graded (but see Chapter 19). Computer aided learning and assessment is a subject in its own right and is not discussed here, but it is worth noting that feedback is provided when wrong answers are given and that records of scores obtained can be made available to the teacher, as noted by Lajoie in Chapter 14.

MULTIPLE-CHOICE QUESTIONS

Multiple-choice questions have become fairly popular in recent years. They are quick to administer and mark/grade, but on the whole are suitable only for short questions requiring a minimum of thought. They cannot probe deeply into understanding or require long answers, though they sometimes ask for a justification of the response. Multiple-choice tests are as much an aide memoire to the student as useful to the teacher and are perhaps most suitable for formative assessment or for quick feedback.

A serious objection to multiple choice is that the distractors might reinforce incorrect ideas. Giving three choices is fairly popular. The student then has a 1 in 3 chance of guessing the correct answer which should be varied in position. A simple right/wrong scoring system might be used. Allowing students to select a "Don't know" option should help reduce guessing and the instruction "Give the MOST CORRECT answer" helps to convey the idea that there is not always a unique correct answer in statistics.

Many multiple-choice questions are bad because there is no attempt to relate them to a context, and it is not clear what is being tested other than an ability to apply a definition and perform a computation correctly. An example is:

Given the probabilities of $x = 1, 2, 3, 4, 5$ being 0.1, 0.4, 0.2, 0.2 and 0.1 respectively:

What is the expected value of x-squared?
a) 5.0   b) 7.84   c) 9.2   d) 6.25   e) 3.03

What is the standard deviation of x?
a) 1.166   b) 1,360   c) 2.8   d) 1.0   e) 1.44

The following questions are better because they test concepts and do not require much computation, but these too can be criticised for their lack of context and realism. An improvement might be to relate them to scaling of examination marks/scores.

The mean of the numbers 5,6, and 7 is 6. If each number is squared the mean becomes:
a) 36   b) greater than 36   c) less than 36

The standard deviation of the numbers 5,6, and 7 is 1.
If 1 is added to each number the standard deviation becomes:
a) 1   b) 2   c) square root of 2

Care has to be taken in setting questions of this type that the answer is not so obvious that the student hardly has to reason in order to get it correct. On the other hand it should be possible to select the answer without doing elaborate calculations. This is not as easy to achieve as it might at first seem. Studies have shown that people are rather poor at "subjective" estimation (Hawkins et al., 1992, Chapter 6).

Multiple-choice questions are used in research studies, for example, to find out misconceptions in order to inform remedial action, or to assess the success of an intervention such as practice simulation exercises. Such questions are not designed for assessment purposes and adverse criticisms of multiple-choice questions do not necessarily apply to them. However, some of these are good models for assessment questions, particularly some of those used in studies relating to probabilistic and statistical concepts. In research it is common to follow up the question with "Why do you say this?" or equivalent.

The following example which is taken from a statistical thinking survey (Swets et al., 1987) is a good example of the kinds of questions asked. The instructions included the statements:

> Please answer each question as well as you can, paying particular attention to the 'Why' part. If you have an idea you are not sure is correct, include it as well. We are especially interested in how you think about these problems.
>
> *If you wanted to investigate whether the Beverly Hills High School students were better or worse in mathematics than the Valley High School students, check which of the following methods you would choose:*
>
> () Pick one student at random from each school, give them the same math test, and compare their scores;
> () Give the same test to every student in the two schools, and compare the total of the scores obtained in each school;
> () Give the same test to every student in the two schools, and compare the highest score obtained in each school;
> () Give the same test to every student in the two schools, and compare the average score obtained in each school;
> Why would you use this method?

Chapter 16 (Wild, Triggs, & Pfannkuch) in this volume focuses in greater detail on multiple-choice questions, their limitations and benefits, and reasonable ways to use them in contexts where other formats are impractical, especially in assessing large college classes.

OPEN-ENDED QUESTIONS

Open-ended questions are a complete contrast to multiple-choice questions and to very specific questions which say what is to be worked out and which lead students into appropriate comments. Probably the most open-ended question which can be devised is of the type "Analyse the following data and report on your findings." The problem with such questions is that as there is no guidance as to what is expected; attempts at them are likely to vary greatly both in quality and in quantity. Students may do every possible analysis imaginable, including some which are inappropriate. It could be argued that questions of this type are mini-projects and could be assessed as such. The assessor needs to decide what analysis it is reasonable to expect the student to do in the time available, and to have a grading system or scoring rubric (see Chapter 3 by Colvin & Vos) which is flexible enough to allow for non-standard and unexpected approaches.

Short open-ended questions involving an element of reasoning can be useful. For example:

What, if anything, is misleading about the following claim?
An increase in sample size will always cause a sample mean to move closer to a true mean.

(Bernklau)

A slightly longer task involving real data is:

An *El Paso Herald-Post* article discussing sports-related deaths stated "Basketball is the country's most lethal sport." This "rather startling finding" was based on 1989-1991 data from the National Center for the Study of Sudden Death in Athletes, which reported the following death numbers: basketball, 70; football, 42; track, 38; baseball, 24; skiing, 16; soccer, 13; weightlifting and hockey, 10 each; wrestling and boxing, 8 apiece; golf, 7. Comment on the conclusions quoted from the article.

(Kaigh)

A more focused task involving analysis of data will ask a specific question and perhaps hint at what is required in the way the information is presented. The following is a good example of how to do this.

A study was conducted on highly intelligent children. The researcher collected data on 30 children on age and IQ. The data are below. The researcher claimed that one student, Dean, scored higher than most students on IQ, but not on age. Is this claim acceptable?

Student	Age (years)	z score (Age)	IQ	z score (IQ)
Ed	12.10	1.07636	143.00	-.86936
Sam	9.40	-1.13647	152.00	1.82864
Dean	13.00	1.81398	148.00	.62953
Ann	9.30	-1.21842	140.00	-1.76869

(note: the original table is much longer than the portion shown here).

(Cohen)

We might well wonder how students who answered "Yes" or "No" without further comment would have been graded! A better wording might be "Perform an appropriate analysis to help determine if he can substantiate his claim."

A question might leave the decision of fine detail to the student but give some guidance as to what is needed. For example, a question which says "Perform an appropriate test to investigate whether group A scored higher on average than group B," where scores are given for members of both groups, is open-ended in that the student is left to choose the test, state the statistical hypotheses, and so on. A question saying "Make two brief comments comparing the given distributions" gives a little guidance as to the kind of response expected and its length. One with a wording such as "Summarise the data in a manner suitable for inclusion in a company report" has an element of open-endedness in it and a touch of realism.

Graham (1990) suggests how to find suitable statistical investigations and how to modify closed tasks to make them more appropriate to the PCAI cycle (Pose the question, Collect the data, Analyse the data, Interpret the results). For example, instead of the closed task "Find and record the number of pupils born in each month of the year," it would be better to ask "Is it true that more babies are born in the Autumn than other seasons?"

Many texts contain case studies, some of which are in the form of open-ended problems. Morris (1989) builds up a case study, presented as letters from Jane, who runs a catering business from home, throughout the book. One letter asks for an explanation of MINITAB output of a customer survey, and the instruction is to write a short report interpreting the information in practical terms, making recommendations for action. In another letter Jane is considering whether to build an extension to her kitchen or to buy or rent other premises and asks how decision theory works. The exercise is to draft a reply using the basic situation and inventing data to illustrate the ideas of decision theory.

These examples show how open-ended questions can go a long way towards meeting the requirements of good assessment questions. They are able to pose "real" questions with "real" data and assess choice of appropriate techniques and the doing and (written) communicating of statistics. They also ask students to explain concepts—a sure test of understanding. However, if there is an obvious link between such a question and the subject matter which has just been covered, the question will not test the student's ability to tackle an unknown problem or to bring together a range of methods learnt throughout a course.

An essay question is another type of open-ended assignment, particularly useful for indirectly exploring understanding. We might ask the student to develop or evaluate a research design, or write about the evolution of statistics, or the use of statistics by the media. A quantitative essay with an emphasis on numbers rather than words is another possibility and might usefully link with the student's other subjects. For topics such as "Is defense a major item of government expenditure?" or "The measurement of poverty," the student would need to find relevant data, present this as tables and/or diagrams, discuss the facts using summary or other measures, and suggest further analysis where appropriate. Giving students examples of papers or books where the mix of words and statistical aspects is about right, such as articles in *Chance* (American Statistical Association), will guide students as to what is expected.

Scoring

Almost certainly the grade awarded to an essay will be based on a general impression—something that arts teachers are used to doing, but a little alien to the typical statistics teacher.

However, essays are a form of assessment well worth considering. They test the ability to organise material from a variety of sources, may test creativity, and almost certainly test understanding. They can be vocational, they allow students to draw on and demonstrate their own knowledge of statistics, and on the whole are not memory dependent.

Regarding scoring of open-ended questions in general, a cross between subjective and objective grading is one way of assessing attempts at open-ended questions. The assessment might be broken down into putting the problem in context, appropriateness of analysis, computational accuracy, and conclusions with each aspect graded on a 5-point scale. Grades could be combined if required. A scoring method developed by Garfield (1993) uses a 0/1/2 scoring with 0 indicating incorrect use, 1 indicating partially correct use, and 2 indicating correct use. The categories she suggests are communication, visual representations, statistical calculations, decision making, interpretation of results, and drawing conclusions. This system adapts well for use in other contexts.

ASSESSMENT OF GROUP WORK

It is important to train students to work in groups because people doing statistical projects need to be able to interact with others; this is a useful life skill in many fields of employment. Sometimes sharing of effort may be the only practical way of completing a problem in the time available. However, the assessment of group work poses many challenges. In general, it involves assessing two separate but related issues: the assessment of the nature and quality of the group *process*, and the assessment of the quality of the *product* of the work of the group. In both cases, it may sometimes be useful to be able to determine the individual contribution of each student to the process or the product, and consider whether each member is given the same grade (i.e., the group grade), or whether adjustments are made depending on the difficulty level of the subtask a student handled, his/her initial skill or knowledge level, and so forth.

As for assessment of process, this topic is handled in some detail by Curcio and Artzt in Chapter 10 of this volume. The product of the group work could be judged holistically, with a score given on a scoring rubric with anywhere between 3-6 levels. Or, separate categories could be established for each aspect or facet of the group product (e.g., problem definition, design/approach, methods of analysis, interpretation, conclusion, presentation), and a weighted score given to each aspect. Often, a way of assessing group work is to award the group a single overall mark based on marks/grades given to the various components of the task performed by the group (see also the scoring rubrics used by Keeler, Chapter 13; or Colvin & Vos, Chapter 3; in this volume). The assumption here is that whether or not every group member has been involved to the same extent with the execution of every aspect of the work, all should have been involved to some extent, for example, in planning and in deciding who would do what. An example of a scoring rubric that can be used in such assessment can be found in Garfield (1993).

Some teachers may want to give a group project a single composite score that takes into account both the quality of the product and some features of the process (e.g., whether the group collaborated well and made effective use of all members' skills and contributions). This is possible but will depend on the amount of information available to the teacher about the group process. In some contexts teachers may be helped by using peer assessment, as discussed below in Implications. One way of assessing the contribution of individual students to a group process involves using a fairly broad grading such as a 5-point scale with a weighting system to take account of the difficulty of the task and (mis)matching between students and tasks. A task that was inherently hard might merit increase of one grade and a task that was too hard for a student might also merit increase of one grade, with corresponding decreases for inherently easy tasks or tasks too easy for a student. There is of course subjectivity in this. Even when we think we are making an objective assessment there is usually an element of subjectivity in there as well.

When students are all doing the same thing but pooling results, each student can be assessed as an individual. If in group discussion about the task there is variation in the quality and quantity of inputs of different students, a mark might be given to reflect this (provided the teacher feels able to make a fair discrimination—not necessarily an easy task!). To some extent a similar problem occurs in any assessment in that some students may be leaders and others followers or sometimes mere copiers, and it is not always possible to detect where credit for originality is due.

In a project type investigation students might work on different aspects; for example, some might concentrate on background reading, some on data collection, others on analysis or report writing. Here assessment poses a real challenge. This is because:

- Tasks may differ in the nature of what is required or the level of difficulty or the time needed to complete them.

- Some students might have tasks on which they are able to perform well, whereas others have tasks they find hard and on which they are not able to demonstrate their highest level of attainment. For example, a student who excels at data analysis but is poor at report writing is likely to get a higher grade if doing analysis than if writing a report.

- The method of assessment appropriate to one task might be inappropriate for another; for example, a fairly "accurate" scoring system might be used for data analysis but a subjective grading system might be better for a report.

- Different stages of the work do not always lead to the same type of output; for example, thinking and planning might lead to no tangible hard copy.

- It may be difficult to sort out individual work from group effort and to assess interaction among group members—an important aspect which should not be neglected in assessment.

In fact it could be argued that awarding individual marks to group members is contrary to the spirit of group work, but sometimes it is important to discriminate between students, for example for the award of scholarships. One way of modifying a group's mark to take account of an individual's contribution is to ask students to rate each other (and even themselves) and devise a weighting for each student based on an average of the ratings (s)he has received including perhaps a rating from the teacher. Again it is difficult to avoid subjectivity but peer assessment is in itself another workplace skill and it does no harm to expose students to it early on. Another option is to use an observation form designed along the lines suggested by Curcio and Artzt in this volume (Chapter 10).

ORAL ASSESSMENT

The relatively new development of oral assessment in statistics is to be applauded because oral presentation of results is as important as written presentation and is something that occurs in the workplace, but there is as yet little experience of it. As with any form of assessment, standardisation of questions and techniques is important, but the administration of oral assessment is more obviously uncontrolled than written assessment, and ideally teachers need some practice in it before using it for summative assessment. A partly subjective broad grading scheme may work fairly well, most likely using a rubric with 4-5 levels, or several categories with an attached weighting scheme.

Oral assessment can be part of a more traditional assessment; for example, students might be asked to clarify written answers or to explain their methods. On occasion it might substitute for other methods. It is useful for assessing young children as when the teacher reads pre-prepared questions and asks students to indicate correct answers on a prompt sheet, for example, to point to the modal group in a histogram. Oral assessment can also be useful for diagnostic purposes to identify points which have been misunderstood or problems experienced by low achievers.

Tape recording or making a video of an oral procedure means that grading can be done independently of the administration of the assessment but could have an inhibiting effect on the students (and perhaps the teacher). Smith and Griffin (1991) describe the use of videos in training and Chapter 14 is also relevant. Negative aspects of oral assessment are that those whose spoken language facility is poor, typically those who are being assessed in other than their mother tongue, those with speech difficulties, and very shy students could be disadvantaged.

IMPLICATIONS

In sum, the key points of this chapter are as follows:

- Setting good assessment tasks can be tricky, but there are many good examples on which to build. Always remember—if you come across a good assessment question keep it carefully!

- Factual knowledge is best assessed through questions where the main aim is to test understanding. Such questions might involve some computation.

- Computational ability can be tested as part of a more challenging problem. We should teach students how to obtain approximate answers but should not necessarily assess them on this.

- Assessment of computer use involves statistical and computing aspects. Assessment of the former is similar to any statistical assessment, of the latter as of any computing tasks, but this is a new area and there are not yet many published examples of it.

- Multiple-choice questions are sometimes useful, but take care not to overdo their use.

- Open-ended questions can be long or short. They can give some guidance as to what the student is expected to do. They are ideal for teaching statistical understanding and skills but grading is partly subjective which the statistics teacher might find a little strange.

- Group work should be encouraged but assessment of it presents new challenges. Little experience exists in this area and there is room for exploration and experimentation.

Many of the traditional methods of assessment in statistics concentrated on rote learning and computational aspects. They are now seen to be unsatisfactory in the new climate of teaching and learning statistics where the understanding of statistical concepts is emphasised. Open-ended tasks, preferably using real data, can be made an important part of the teaching process and are valuable learning tools, but marking and grading students' attempts at such tasks is a new area for many teachers. Multiple-choice questions and questions where the method is prescribed are less satisfactory in assessment, but used with care and in moderation can be useful—and grading does tend to be easy.

The increasing use of computers, and of group and oral work in statistics is to be applauded, but their use in assessment is a relatively unexplored area, and the development of suitable assessment tools with associated guides to grading is required. The practising teacher has an important role in contributing to the growing body of knowledge about these methods.

It is an exciting time as regards methods of assessment. Don't be afraid to experiment in your use and development of assessment tasks and grading methods—be imaginative and innovative. Use a variety of methods for a richer insight into student learning. Above all be critical of your own and others' assessment tasks. Note what works well and what does not and share your findings.

Most of this chapter has been concerned with assessment by the teacher, but the students themselves could mark their own or others' attempts at some of the more routine assessment tasks if provided with suitable "model" answers. In the case of formative assessment this could be quite satisfactory. Making random checks on the process would help ensure students were honest.

Part of the purpose of assessment is to monitor progress and to take appropriate action if students are not performing satisfactorily. An interesting possibility is to get students to use quality control techniques to monitor their own progress. As higher and lower control limits each student could take the highest and lowest scores (s)he expects to obtain on assigned work. More details and a pro forma are given in Kimmel (1992).

The context in which statistics is presented is important. If students do not care for the context they will not be motivated to learn and in consequence will perform badly on assessments. We need to be careful that we do not inadvertently form a wrong opinion of the statistical expertise because the student has been turned off by the context; for example, research has shown that females perform better on "people" questions than on male oriented or abstract questions (e.g., Clark, 1994). Many recent texts use real data or base questions on actual studies. This is likely to motivate students but the teacher may need to be selective as regards the emphasis in the real data taken from texts and other sources for teaching and assessment purposes. In particular these should be gender-free and should not be biased for or against any particular ethnic or social group.

These ideas are complemented by other chapters in this volume, especially those discussing "alternative" assessment, such as Colvin and Vos, Keeler, and Lesh. But more research needs to be done, both in traditional "academic" form (e.g., about assessment with computers), and more importantly, in the form of "teacher research" or "local research." In these latter forms teaching teams or departments interested in revamping their assessment schemes may try to selectively implement new forms of assessment and see what is the value of the new information obtained for informing teaching and giving feedback to students, or what is the impact on students, e.g., of using certain scoring rubrics, or giving some credit to group process, not only to group product.

Acknowledgments

Thanks to David Bernklau of Long Island University, Steve Cohen of Tufts University, Mike Fuller of the University of Kent at Canterbury (UK), and W. D. Kaigh of the University of Texas at El Paso for their willingness to make their assessment questions available for inclusion in this chapter.

Chapter 16

Assessment on a Budget: Using Traditional Methods Imaginatively

Chris Wild, Chris Triggs and Maxine Pfannkuch

Purpose

Because many of the authentic assessment methods described in this book tend to be very demanding of teacher-time, there is still an important place for finding ways to employ inexpensive, traditional methods more creatively in an attempt to come closer to achieving the same goals that those who advocate authentic assessment methods are targeting. The basic idea is to identify the elements of statistical thinking that we want to foster and then find ways of testing these elements with objective assessment methods (in particular, multiple choice). This chapter explores the extent to which objective testing can approximate the results of authentic assessment techniques and the extent to which it falls short. We provide guidelines for the writing of objective test items, together with examples.

INTRODUCTION

The authors are part of a team teaching introductory statistics to 2,600 students (200-300 per section) in a research university on a tight budget. Although the first-year statistics course that we teach was once primarily concerned with statistical calculations, we have now become more concerned with producing "intelligent citizens" than specialist statisticians. The majority of our students will not become statisticians, but all will be users or consumers of statistics in their careers and daily lives. We live in a world where, more and more, machines do calculations and data processing. The essential human role is to ask questions, and to obtain, interpret, and synthesise information from a variety of sources. We want to help our students to prepare themselves for this new world. We want to embed our teaching in the rich context of the investigative process, to foster the abilities to carry out investigations, to critique the investigations of others, and to communicate what data are saying. Thus, we are striving for the same general goals as those presented in the first part of this book. The most powerful signal that we have for telling students what we believe to be important is assessment.

In our university, there are pressures to improve teaching as well as increase the research output of faculty, but the main institutional focus is on the latter. With continued downward pressures on the funding of higher education across the English-speaking world, more and more teachers of statistics seem fated to have to do more with less as faculty/student ratios erode and classes grow ever larger. Despite the pressure for increased "efficiency," we are striving to pursue the same goals as those who advocate authentic assessment methods in statistical education. However, because many of the assessment methods in this book tend to be very demanding of teacher time, our challenge has been to describe our search for ways to employ inexpensive, traditional methods more creatively in an attempt to pursue the same ends.

The basic methods that we use—assignment work and multiple-choice testing—are probably an anathema to most contributors to this book and yet the time savings, when applied to 2,600 students, are overwhelming. The focus of this chapter is multiple-choice testing. Wild, Triggs & Pfannkuch (1994), an earlier revision of this chapter, contains a fuller discussion including the strategies we apply for assignment work to remedy the deficiencies of multiple choice. Although multiple choice cannot do at all well some of the things we value most, we have been pleasantly surprised by what we have been able to achieve and at some of the positive side effects on our teaching. This chapter explores some strategies for making the thinking required in a forced-choice test much more like that required in a real statistical investigation, and it explores the pros and cons of multiple-choice testing using a sample set of questions. Many of the suggested strategies apply just as much to free-response testing as they do to forced-choice testing.

This chapter is structured as follows. The following section, entitled "The Assessment Challenge in Large Classes," outlines the principles that we believe in with regard to assessment and the reasons that led us to adopt forced-choice testing techniques, and goes on to introduce the authors' techniques for overcoming some of the difficulties. This is followed by a section called "Practical Examples," which contains a suite of questions that illustrate the use of these techniques. All this becomes background to the deeper discussions in the following section—"What Have We Learned?"—which has subsections entitled: "Costs"; "Educational advantages"; "Considering some criticisms"; and "What multiple choice cannot do." The chapter closes with "Implications".

THE ASSESSMENT CHALLENGE IN LARGE CLASSES

Principles guiding assessment

The focus for assessment has traditionally been on certification, that is, as a measurement device to enable teachers to certify that a student has a certain level of mastery of ideas/material. More recently, the roles of assessment in motivating, focusing, and monitoring the learning process, and monitoring the interaction between teaching and learning, have been receiving more emphasis. Fortunately, these roles are complementary. Whether we like it or not, the assessment materials we give to present students and have given to past students strongly influence where students put their efforts. To a very large extent, what you test is what you get. Assessment that guides student attention in less than optimal directions fails in the certification role. Finding holes in student learning for certification purposes also exposes areas where teaching could be improved. Curricular goals should be formulated to meet the needs of the "customers" of the course, e.g., the students themselves, employers, society at large and teachers of future courses (Wild, 1995). Assessment should endeavour to assess the extent to which these goals are being met. Because no form of assessment is perfect it is better to test the important inadequately than it is to test the unimportant well. In addition, assessment must be manageable within the budget (of people, money and time), and be perceived to be fair by students. General considerations about goals and assessment are considered in the first part of this book.

Constraints

Our first-year course runs through a full academic year. We have 2,600 students who are taught and assessed by 12 lecturers who should each be allotting less than 25% of their time to this course. In some years, as few as two of the lecturers are experienced as both statisticians and teachers. Most of the rest are Ph.D. students who are relatively inexperienced in both roles, although we usually have one or two who are experienced as teachers but not as statisticians. Backup tutorial assistance is provided predominantly by Masters students, and there is enough marking (grading) assistance (mainly undergraduate) for about 9 assignments, each of which take the students about 3 hours to complete.

Our switch to multiple-choice testing was entirely a resource decision, a way of saving time. Our assessment challenge is to get as close as we can to the educational goals we have outlined above, to emphasise statistical thinking and not just routine mechanical tasks using the tools at hand. But it has not all been bad; there have been some quite positive side effects that will be discussed later. Putting additional effort into assessment will necessarily result in something else suffering because almost all the people on the teaching team are already committing far more time to their teaching than is good for their careers or for the fulfilling of institutional priorities. We are not unique in this, for no one lives in a world of abundant resources. Everyone is in the business of trying to put a limited set of resources into the places where they will do most good, with all the uncertainty and under-informed tradeoffs that implies.

Our stratagems

We cannot do everything we want to do with multiple-choice methods and need other forms of assessment as well. Philosophies underlying the assignment work that complements our tests and examinations are discussed in Wild, Triggs & Pfannkuch (1994). However, we have found that we can overcome some of the inadequacies that are often claimed to be inherent to multiple-choice testing. We can do something about incorporating synthesis and evaluation, about minimising rote learning as a primary study focus, about testing big ideas rather than merely subtleties, and about interrelationships between ideas. We can also do something about simulating the stages of an actual analysis of data.

We use the following stratagems to accomplish these tasks:

1. Begin by collecting a file of real stories/data sets.

We do not think about types of questions at this stage except in the broadest of terms. Ideally, these data sets are so sufficiently rich in context and features that many questions can be built around a single story. Although we have only presented 8 questions about the data set in the next section, we could easily have asked 30. We would typically have five or six data sets in a three-hour examination, although not all of them would be as elaborate as the one here. (We have never found it necessary to use artificial "data" and would view doing so as failure.)

2. Present the background, data, and a fairly large array of numerical and graphical summaries derived from it as a complete package.

All the information is presented together in preference to releasing the information in small pieces with the questions as they require it. This strategy forces students to synthesise and make use of a body of information. We believe in supplying more information than will be used so that students have to become selective. Most of the questions in the "Practical Examples" section (to follow) come from an actual examination.

3. Try to ask as many questions from the same story as possible, letting the story/data/situation suggest the questions.

Some reasons for this are as follows. First, we are more likely to get some real statistical thinking into the exam paper. Second, it is a way of triggering new ideas and fresh questions. Third, with only a few stories we can afford more complexity, and students can afford to make the time investment necessary to understand the context. Having fewer stories also shortens the effective paper length.

4. Include questions of the following five main types:

We have identified 5 aspects of the statistical process and statistical understanding that assessment should address. These are: (i) critiquing practical aspects of studies; (ii) interpreting data-information and making inferences from it; (iii) interpreting and understanding statistical ideas; (iv) specifying what techniques should be used in a given situation; and (v) performing mechanical tasks (e.g., calculations).

5. Break down tasks into subtasks and examine each subtask (alternatively, a coherent set of ideas) separately.

We do this in part because we cannot give partial credit, but there are more important reasons. This division of tasks reduces the gap between the answer to a question and the thinking processes used in its construction—the only information the teacher gets is the answer. If several skills are mixed up in a question, then if students get it wrong we do not know where the problem lies. If the question does not have a coherent theme, it will not be useful as a means of focusing the attention of future students on important points. Thoughtful question writing that tries to get at thinking processes can motivate us to isolate exactly what it is that we are trying to teach. Unless routine arithmetic is being tested, all necessary summary statistics (and a few plausible distractors) are given. Where answers are numerical, most alternative answers are obtained by using the types of mistakes students commonly make in written work. With verbal concepts, we try to use common misconceptions.

6. Provide other information to cut down on unproductive rote learning.

There are many ways of doing this, from conducting open-book examinations to allowing crib sheets. Our examination papers incorporate almost all of the formulae that students will need to use. We try to make conscious decisions about what is worth committing to memory and what is not. We make almost no demands upon students to remember formulae.

7. For questions about interpreting statistical concepts, or for "what to use where?" questions, bury one false statement amongst a collection of true statements.

To examine the understanding and interpretation of statistical concepts, and knowledge of when and where particular tools are appropriate and useful, we construct questions which ask students to identify the one false statement in a set of five where one statement is false and the other four statements are true. The true statements are usually important statistical messages, concisely stated, that we wish to transmit to future students. This reinforces the teaching and minimises the amount of random misinformation being fed into the brains of future students when they use old tests in their exam preparation. The correct answer can be obtained either by recognition or by eliminating the other possibilities. If students are being asked to critique something in particular, e.g., a particular plot, or anything to do with a particular data set, then one also has the option of asking students to identify the true statement among a set of false statements. Because this type of information is transient, it does not matter if much of the information presented is false. See Frith & MacIntosh (1984) for other ways of constructing multiple-choice tests out of true and false statements.

8. Avoid statistical jargon in interpretational questions.

The statements used in interpretation questions should be expressed in language that is as close as possible to plain English. Why? Because we believe that the cycle of real statistical investigation begins with questions that are intuitively understandable and nontechnical and has not been successfully concluded until conclusions are understood on this same level (Wild, 1994). An important component of the ability to communicate statistics is to be able to interpret the results of abstract manipulations in terms of concrete reality. It is all too easy to work successfully in an abstract framework with no real understanding of what it all means. Thus, we have to take care that the language used is no more complex than it needs to be. Statements about statistical ideas that are concise, correct, and easy to understand can be extremely difficult to write.

We believe that most of these eight recommendations apply much more generally than just to multiple-choice testing. We would want to use every stratagem discussed above (except for stratagem 5 and stratagem 7) for a time-limited free-response test. In a free-response examination, the intent of stratagem 3 (asking a multitude of questions) may be approached from the opposite direction by asking one or two totally open-ended questions and allowing students to dig around in the available information and reach whatever conclusions they can from it. Without signposts to stimulate thinking in the host of directions possible, one may get some very blank answer books because students have had little opportunity to display what they do know. However, the ability to handle such open-ended questions is obviously what we should be aiming for.

PRACTICAL EXAMPLES

Most of the questions which follow come from the last examination that the authors worked on together. They exemplify many of the strategies described above. The original examination paper also included traditional items requiring using numerical information that had been provided. These skills are still important and still tested, but no examples are presented here. The questions that follow are based upon research about measuring human genetic damage resulting from environmental exposures (Margolin, 1988). It comes from real research (stratagem 1) and was presented as a block (stratagem 2). Only 8 sample questions are posed, but in practice, we would keep generating questions from this data until we ran out of ideas (stratagem 3).

The information provided to students as a "cover story" on which sample test items focus is shown in Figure 1. We should note that the text is one we use for college-level students already proficient in academic-style reading. Texts with similar content but simplified language can be adapted from newspapers and magazines and modified for use with students who have a more limited reading background. Students received five additional tables containing the raw data, numerical summaries, and the results of analyses (see Wild, Triggs & Pfannkuch (1994)).

Background information

Monitoring for human exposure to environmental agents that might cause genetic damage and determining the type and extent of such damage is a subject that is commanding increasing attention. One method suggested as a possible measure of the genetic damage present in an individual's DNA is derived from counting the number of SCE's (sister chromatid exchanges) observed per cell. An SCE results from a reciprocal exchange of DNA between two spiral filaments that constitute a chromosome (sister chromatids). For each individual, a sample of cells is taken and the average of the resulting SCE measurements is recorded. This gives the value of the variable we call MSCE for that individual.

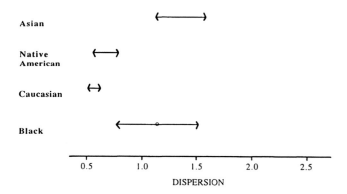

Dot plots for data on DISPERSION with superimposed LSD-display intervals.

One scientist looked at the average MSCE levels for four different racial groups and found no significant difference. However, other scientists have suggested that a measure of the variability of SCE measurements within an individual compared to that individual's average SCE level (giving a variable we call DISPERSION) may be a better measure of genetic damage for that individual.

Figure 1. Cover story for test items

Example 1: What to investigate next?

Suppose that you are listening to a conversation between five students about this problem. Which student would you agree with **least**?

(1) There appears to be a significant difference in DISPERSION levels between "Asians" and "Caucasians." I would like to check out whether this was due to a difference in the time of the cell scoring.
(2) We are not told about the genders of the subjects. I would like to check out whether the differences could be due to gender differences rather than racial differences.
(3) Are the smoking histories of the groups comparable? I would like to check whether differences in exposure to smoking explain the differences between racial groups.
(4) If you disregard the outlier in the "Black" group, only the "Asian" group stands out as different from the others. I would like to check out whether the "Asian" group are first-generation Americans as diet may be a possible cause of the difference.
(5) There being no real genetic differences between racial groups, I would like to check out whether these published differences result from the political beliefs of the researchers.

The habit of looking at a study and posing questions for further exploration is a crucial part of statistical thinking (Wild, 1994) and one that we really need to emphasise for future students. It is also an enormously difficult area to address with multiple choice. We appeal to our earlier adage, "Better to assess the important inadequately than the unimportant well." We do not expect this question to work well as assessment per se. Its main role is to say to future students using the exam in their studies, "Hey, this is important. Don't overlook it." It falls short in that we can only present ideas to be validated. We cannot ask students to come up with their own ideas. Our answer: (5).

Example 2: Nonsignificant versus no different

In this particular study and ten other similar studies, no significant difference was found between the sexes. Recently a researcher did find a significant sex difference in SCE levels and had her findings challenged because "everyone knew" a difference did not exist.
Which one of the following statements is **false**?

(1) The researcher who found a significant difference could not have used randomly sampled experimental subjects.
(2) On the basis of the previous studies, "everyone" had erroneously accepted the null hypothesis of no difference between the sexes.
(3) A real but small SCE sex effect would often not be detected by studies with small sample sizes.
(4) In the long run, the practice of "rejecting" hypotheses at the 5% level of significance results in 5% of studies investigating differences where none are

present, coming to erroneous conclusions.
(5) Misunderstandings would have been less likely to arise if each of the studies had quoted confidence intervals along with the *P*-values.

We have taken an incident requiring statistical judgement and run items that may be causes or implications of the incident. This follows a pattern we use of confronting students with common misconceptions that arise in statistical thinking as a means of developing a critical attitude towards, and an understanding of, statistical aspects of published research papers. Most of this question attempts to put the student in the position of a research reviewer, to conjecture why and where possible errors of reasoning could have resulted in a "common belief" by scientists. We have asked for one false statement to be picked out from a batch of true statements (stratagem 7) with the true statements reinforcing important messages so that they act, at the same time, as a learning device. This question draws on more ideas than we would generally like in one question (stratagem 5), but this priority is overridden by the desire to avoid obviously stupid alternatives and stay close to factors at work in the incident. It works at the interface between three of the aspects listed under stratagem 4, namely: critiquing practical aspects of a study, interpreting data, and interpreting statistical ideas. (The multiple-choice format for communicating an answer to this question is imperfect in that no explanation of why the false statement is false is required.) Our answer : (1).

Example 3: How do we look for a relationship?

If we are interested in looking at the relationship between a person's MSCE value and their DISPERSION value, we should look at:

(1) A scatterplot of MSCE versus DISPERSION.
(2) A stem-and-leaf plot of the differences between MSCE and DISPERSION.
(3) Side by side stem-and-leaf plots of MSCE and DISPERSION.
(4) A stem-and-leaf plot of the ratio, MSCE divided by DISPERSION.
(5) Side by side boxplots of MSCE and DISPERSION.

This question tests which plotting techniques give the desired information for a given situation. When data is produced it is important to know how to deal with that data and what techniques are appropriate for the formulated question (see stratagem 4). This question is much more focused than the previous one (per stratagem 5). Similar questions can be used to address the choice of modelling techniques. Our answer : (1).

Example 4: Reading and interpreting the plot

Which one of the following statements about the data on the variable DISPERSION (see Fig. 1) is **false**?

(1) Removing the observation with value 2.63 from the "Black" group would increase the evidence of a difference in mean DISPERSION between the "Black" group and the "Caucasian" group.
(2) The highest average level of DISPERSION appears to be in the "Asian" group.
(3) There is an outlier in the "Black" group.

(4) Removing the observation with value 2.63 from the "Black" group would strengthen any evidence of a difference in mean DISPERSION between the "Black" group and the "Asian" group.
(5) Ignoring the observation with value 2.63 in the "Black" group, the variability of the DISPERSION variable appears similar for the "Black," "Caucasian," and "Native American" groups, but looks somewhat larger in the "Asian" group.

In the list given under stratagem 4, this question corresponds to "interpreting data information." Students have to know that the features being described here are best summarised in the plots of the raw data—no reference clues are given. Statistical jargon has been avoided to the best of our ability (stratagem 8). We are after an accurate reading of information in the plot and an intuitive understanding of how point patterns relate to formal evidence of differences between group means. We would often ask about features of plots without the link to formal inference being made here; that link was suggested by the data itself. An appreciation and intuitive understanding of variability is also reinforced. We have used a single false statement again, (per stratagem 7), partly because writing true statements about a plot like this is much easier than coming up with false but plausible statements, and partly because the four true statements used will help future students learn how to "read" data like this. An inadequacy of the question is that students are validating reactions to the plot rather than generating their own. Our answer : (1).

Example 5: Verbal understanding of a technique

Which one of the following statements about the application of one-way analysis of variance to the data on DISPERSION is **false**?

(1) The null hypothesis being tested is that all of the underlying mean DISPERSION values for all groups are identical.
(2) The F-statistic compares the variability between the sample means with the variability within samples.
(3) Large values of the F-statistic provide evidence of a difference between underlying true means.
(4) If the P-value is large, this implies that all the means are the same.
(5) A small P-value suggests that the data provides evidence of differences in underlying mean DISPERSION between some of the groups.

In the list given under stratagem 4, this question corresponds to "interpreting statistical ideas." It tackles the big ideas of one-way analysis of variance and the F-test. It combines ideas about the hypothesis tested and the meaning of large and small P-values in the ANOVA setting. Sometimes we would take two questions to do this. The P-value as a concept would have already been tested on its own in a simpler setting. This question is about what is special about one-way analysis of variance. The understanding of how a significance test is set up statistically, the interpretation of summary statistics, and what can be concluded from them are tested here. Jargon words are translated into terms that are as concrete as we could get them (stratagem 8). We also rely on stratagem 7 (all statements but one are true) to reinforce the main ideas about one-way analysis of variance. Our answer: (4).

Example 6: The purposes of formal tools

Which one of the following statements about analysing data is **false**?

(1) Plots can help tell us which methods of analysis are appropriate.
(2) It is good general practice to look at plots of data before performing formal analyses.
(3) We may use formal tests and intervals to help determine whether effects we have seen in our data are real or may just be due to sampling error.
(4) Even though we test null hypotheses of the form $H_0: \mu_1=\mu_2$, we would seldom, if ever, expect two population means to be identical.
(5) A P-value for testing a null hypothesis decided upon after looking at the data provides at least as much evidence against the hypothesis as it would have if the hypothesis had been decided upon before the study began.

In the list given under stratagem 4, this question corresponds to "what techniques should be used." It is a general question about approaches to data analysis. It seeks to reinforce problem solving strategies and what can and cannot be reasonably concluded (stratagem 7). Our answer: (5).

Example 7: How is the P-value obtained?

If the *t*-test statistic for testing for a difference in mean MSCE level between "Native Americans" and "Caucasians" is 1.7, then the (2-sided) P-value for this significance test is:
(1) $pr(2\ T \geq 1.7)$ (2) $pr(T \geq 1.7)$ (3) $pr(T \geq 2 \text{ ¥ } 1.7)$
(4) $pr(T \geq 1.7)/2$ (5) $2 \text{ ¥ } pr(T \geq 1.7)$

This question examines the relationship between the test statistic and the 2-sided *P*-value in a non-traditional way. Other variants for testing the relationship between alternative hypothesis, test statistic, and *P*-value range from simply referring to a table to identifying shaded pictures. Our answer: (5).

Example 8: Interpreting a particular confidence interval

Suppose that we are calculating a 95% confidence interval for a difference in mean DISPERSION level between the "Blacks" and "Native Americans." Which one of the following statements is **true**?

(1) If the interval is (-0.1, 0.9), then with 95% confidence, the underlying mean DISPERSION for "Blacks" is somewhere between being 0.1 units bigger and 0.9 units smaller than the mean DISPERSION for "Native Americans."
(2) If the interval is (-0.1, 0.9), there is a significant difference in the means at the 5% level.
(3) If the interval is (0.1, 0.9), then with 95% confidence the underlying mean DISPERSION for "Blacks" is bigger than the mean DISPERSION for "Native

Americans" by between 0.1 and 0.9.
(4) If the interval is (-0.9, -0.1), then there is no significant difference between the means at the 5% level.
(5) If the interval is (-0.1, 0.9), then with 95% confidence the underlying mean DISPERSION for "Blacks" may be smaller than the mean DISPERSION for "Native Americans" by more than 0.1 units.

In the list given under stratagem 4, this question acts at the interface between "interpreting data" and "interpreting statistical ideas." The statistician has "analysed" the data and come up with a confidence interval, but what do these numbers mean? This question examines interpretation of the numerical values of the confidence limits for a difference between two means (in contrast to another question which would target what confidence intervals mean as a general concept). Communicating statistics by being able to translate statistical ideas into plain English is a goal of the course. Three of the statements manage to obey stratagem 8 (nontechnical language), except for the jargon phrase, "with 95% confidence." Two statements make the link between confidence intervals and significance testing. We can thus expect some mixed messages in the item analysis for this question. Our answer: (3).

WHAT HAVE WE LEARNED?

Costs

The biggest advantage of forced-choice testing methods is that when a single test is given to a large number of students, faculty time is saved because the grading costs are negligible compared with other methods of assessment. A 65-question multiple choice paper takes us about 6 person-weeks to produce (including a rigorous checking process). A significant amount of that time is dictated by our desire to use fresh, context-rich data sets, which we would want to do irrespective of the form of examination. If our stratagems are followed, once the data file has been assembled most of the questions almost write themselves. We would put the additional cost of setting the examination in multiple-choice, as opposed to free-response and producing marking guides, etc., to be about 3 person-weeks and certainly no more than four. However, because of our large numbers of students, this 3-week fixed cost is overwhelmed by huge time savings on marking. It used to take us about 25 minutes to mark a traditional 3-hour examination script at this level (at that time we examined only mechanical tasks — verbal responses would take considerably longer to mark). Using the 25 minute figure, we estimate the marking costs for a traditional examination for our system to be at least 27 person-weeks, 9 times the cost of setting the exam in multiple-choice format. The additional cost is equivalent to one person committing 50% of her or his entire year to marking! We emphasise that the comparison here is just between multiple choice and another relatively inexpensive method, namely, the traditional 3-hour free-response examination.

Educational advantages

Other recognised advantages (Popham, 1990; Gronlund, 1993) are that a good deal of subject matter can be covered in a reasonably short period of time, and that marking is objective and consistent. We have found that the decoupling of skills forced upon question writers by the multiple-choice format has had some unexpectedly useful side effects.

The targeting of single ideas. Multiple-choice tests can form an information-rich source that is excellent for highlighting important ideas for future students. Having taught an idea or set of ideas, one can point very directly to that idea being assessed in the past.

Practising those parts of a skill which are causing problems. The breaking down of tasks into sub-tasks to construct multiple-choice questions can enable students to see which particular skills or parts of a skill are causing them problems and provide a means by which a student can devote additional effort to those troublesome areas alone.

Encouraging reflection and discussion about teaching. An advantage of the multiple-choice method that took us completely by surprise was that the discipline imposed by the format on question writers can improve teaching. Having to decouple skills and think through a whole series of options to write alternative answers for multiple-choice questions forced us to confront many of our own underlying assumptions and analyse in much more detail what it was that we really wanted to teach. Multiple choice alternatives tend to lay these things out for the whole world to see. Trying to write alternatives that assess big statistical ideas has increased the pressure on us to come up with ways of expressing those ideas concisely in plain English. Everyone on the team discusses and criticises each question *and all alternative answers*. The whole assessment process is completely explicit. Differences between teachers in preconceptions and interpretations are discussed at length with positive benefits for future teaching and future assessment. Any other forced-choice method can have these benefits.

Confronting students with common misconceptions. This is useful for refining the understanding of concepts. Once its value is appreciated, it can be done in any test format, but the nature of multiple-choice question writing forces teachers to think this way.

Complete consistency in marking. We get better quality control from a multiple-choice method than we used to get with long-answer questions. With long-answer questions, the marking scheme for a question tended to be read only by the person who made it up and then used it. The myriad of small decisions still to be made when marking a free-response question reflect the opinions of a single person. The credit the teacher should give for the thinking a student has done after taking a wrong turn (such as a simple computing error) on a long-answer question poses major problems, both in terms of equity and consistency. The big variation between different markers grading the same long-answer questions is well known, and there is even appreciable within-marker variation. Yet, marker variability pales into total insignificance when compared with the additional variability introduced by the practice of giving completely different tests to the students in different sections (streams) of a course. The "fairness" issues here should not be taken lightly. Students often end up in a particular section, or class, through a sequence of random events related to rooms and timetables—it is not a matter of informed choice. Therefore, they should not be penalised for ending up in the "wrong," or difficult, section.

Some practical benefits for the exam itself. We used to find that with long-answer questions that yield partial credit, a single idea could accumulate credit across different questions throughout a paper. With multiple choice, we give one unit of credit in one place for one idea. With long-answer questions, mistakes or gaps in knowledge in early parts of a question can badly damage answers to the rest of a question.

Considering some of the criticisms

The criticisms made of multiple-choice testing are all valid to a greater or a lesser extent. This section moderates some of the criticism and outlines some additional attempts to mitigate the deficiencies.

"High level thinking." The idea that multiple-choice methods cannot test "high-level thinking" and is confined to low-level skills such as recall and application is overly pessimistic (Pandey, 1990). There is a great deal more analysis, synthesis, and evaluation required by the examples we have presented here than in many free-response items written by people who do not have our commitment to exploiting context-rich real data. It could be done better with free responses, but not without cost, particularly for those students who were stymied early in the process and thus lost the ability to show what else they knew or could do.

Entrapment by subtleties and fractionating knowledge. The nature of writing alternative answers can encourage a puzzle-writer's mentality whereby one ends up trying to trip students up with subtleties rather than examining big ideas. Because everything has to be broken down into small pieces, it can be easy to fractionate knowledge at the expense of interrelationships between concepts. However, we believe that these are pitfalls that question writers need to be aware of, and try to avoid, rather than being inherent flaws of the format. The statements we have used to construct examples 2, 4, 5, and 6, in particular, draw heavily on relationships between concepts.

"Multiple choice encourages rote learning." This is by no means inevitable. Even our mechanical questions (not presented to conserve space) rely very little on rote-memory. Teachers have to make conscious decisions about what is worth committing to memory and what can be obtained from reference materials when it is required. They must then provide the support materials that permit students to work in this way. We expect recall of key ideas only.

Tests used as a study resource. Because multiple-choice tests are so rich in information, we have found that some students use past tests as their primary study resource in preparing for future tests. This underlines how vital it is that test questions be focused on the most important parts of the course and not on obscurities.

Deleterious effects of time constraints. Criticisms such as "you only have two or three minutes per question" are oversimplifications. We gain from students only having to master one story to answer a large number of questions. The answers to some questions come almost immediately, leaving much more time for items that require more reflection. As with any other form of time-restricted test, it is important not to pack too many questions into an exam paper. We would far rather have students finish early than have them under severe time pressure, for such measures more accurately their in-depth knowledge. To deal with even stronger criticisms, we discuss areas where multiple-choice testing is perceived as totally deficient in the next three items.

The ability to perform a sequence of tasks unprompted. Multiple choice is clearly deficient here. We have to break down tasks into a series of subtasks to be examined separately. At each stage only the next step of logic, or a process, needs to be supplied, and then it need only be recognised. For largely mechanical tasks, it is often not practically important for later life to remember the sequence of steps precisely. One can look up the steps if and when one needs to use them. If they are used often enough, recall will soon follow.

Mechanical procedures are also prime candidates for automation. Take for example significance testing: The essentially human inputs, knowing what to test, being able to interpret the results of the calculations in terms of the original problem, and awareness of the need to check assumptions (e.g., about the distribution being normal) can be targeted using multiple choice. The graphical questions in our examples test accuracy of interpretation. This does fall short of what we would prefer: "What does this display tell you about the data?"

Language and communication. When responding to multiple-choice questions, students are forced to conform to the examiner's language in the formulation of the answer as well as the formulation of a question. Three aspects to this issue should be discussed.

- Fairness: Multiple choice is no less "fair" than free response. It is true that despite having its language polished by a whole team of people who are intent on minimising the possibility of misunderstanding, many alternative answers will be misunderstood by some students, and a small proportion will be misunderstood by large numbers of students. *All communication is imperfect.* Free response leads to unpolished one-to-one communications and therefore, in all likelihood, a greater probability of communication failure. A marker (grader) may cope with this by assuming that, whenever an answer fails to communicate itself to her or him, it is the student's fault. This is scarcely "fairer."
- Learning to communicate: The big problem with putting words into someone's mouth, as forced-choice testing does, is that it does little to help students develop their own voices. The best we can do with multiple-choice questions is to try to ensure that students have the ideas right and are used to seeing these ideas in simple language. This is one of the flaws of multiple choice that we pursue in the next subsection.
- Lack of information on student thinking: Pfannkuch has been researching the reasoning processes students use to form their answers by using extended, open-ended interviews of volunteers. Her research has shown that students go through very rich reasoning processes as they consider alternative answers and that these processes can display a great deal of statistical thinking. With forced-choice testing, one only sees the final choice. The greater the amount of reasoning needed between question and answer, the more severe the problem. This situation is much easier to circumvent with mechanical questions than it is with interpretation questions (simply break the task down into smaller components). This, of course, in no way obviates the need for the interpretation questions. Our hope is that our research into the thought processes of students may suggest new items that get closer to the ways students think, ways which can help steer that thinking in productive directions.

The random element. A much more obvious disadvantage of forced-choice testing is the random element introduced by students guessing answers. This problem becomes less serious as the number of questions becomes large and as the number of distractors becomes larger (we use 5 rather than 3 or 4—it is too hard to write more plausible answers).

What multiple choice cannot do

There are crucial elements of the statistical process that are impossible to experience and assess with forced-choice testing; in particular, anything requiring open-ended thinking. This often shows up in problem formulation (or question generation), in developing the ability to communicate, and in developing the ability to work or reason through substantial problems without step-by-step prompting. Forced-choice testing ties one into a world where everything is either right or wrong. There is no room for coming up with competing strategies and different perspectives. (Essentially the best one can do is to ask whether this is a promising approach or is it not, and is this one of the advantages/disadvantages of a certain choice or is it not.) More complex forced-choice test items such as assertion/reason items can do very slightly better (Frith & MacIntosh, 1984).

IMPLICATIONS

The authors have worked together on a team that teaches an introductory statistics course where there is more emphasis on fostering statistical thinking than on producing people who can perform a limited range of calculations expertly, and in which there is no emphasis on inculcating the ability to construct a mathematical derivation. Where the availability of resources and suitable coaches permits, we believe in the apprenticeship model of statistical education—students working on real problems from beginning to end under supervision—see the discussion of Romberg, Zarinnia & Collis (1990) on instruments of assessment and Wild, Triggs & Pfannkuch (1994) on assignment work (including a strategy for mimicking the investigative cycle within the confines of the traditional assignment and on strategies for enhancing communication).

Sheer weight of student numbers and a poor staffing ratio compel us to retain the traditional methods of forced-choice testing and regular assignments marked by student markers. Our challenge is to push the limits of these traditional methods and come as close as is possible to achieving the goals pursued by authentic assessment techniques described elsewhere in the book.

The focus of this chapter on traditional assessment techniques was multiple-choice testing. The following strategies were presented, illustrated, and discussed in detail:

1. Begin by collecting a file of real, context-rich data sets and stories.
2. Present the background, data, and a fairly large array of numerical and graphical summaries derived from it as a complete package.
3. Ask as many questions from the same story as possible, letting the story/data/situation suggest the questions.
4. Incorporate questions: critiquing practical aspects of the research; interpreting data-information and making inferences from it; interpreting and understanding statistical ideas; specifying what techniques should be used in a given situation; and performing mechanical tasks (e.g., calculations).
5. Break down tasks into subtasks and examine each subtask (alternatively, a coherent set of ideas) separately.
6. Provide enough information to prevent students wasting time on unproductive rote learning.
7. For questions about interpreting statistical concepts, or for "what to use where?" questions, bury one false statement amongst a collection of true statements.
8. Avoid statistical jargon in interpretational questions.

Only stratagems 5 and 7 are peculiar to forced-choice testing and only 7 to multiple-choice. The rest apply to almost any form of formal test, whether forced-choice or free-response, and to any level of the school curriculum. The above strategies are simply devices for getting as close to the reality of a statistical investigation as we can in the artificial context of a time-limited test.

Multiple choice does not allow students to experience some crucial parts of a statistical investigation and must be supplemented with other assessment methods. When circumstances dictate the heavy use of multiple choice, one should not simply accept the popular perception that it is an obstacle to better learning, but instead welcome the challenge to find creative ways to assess the skills that really matter. The disciplines imposed by the multiple-choice format on those teachers who are serious about trying to write good questions can have very positive effects on the interaction between teachers and on their teaching. The necessity to break down skills and ideas into component parts forces a deeper consideration of what it is that one is trying to teach. Disagreements between teachers over the meanings, usefulness, and correctness of questions and alternative answers provide profitable learning experiences for all involved. Whatever we do, we will always see students who are "collectors" who only learn isolated facts, "technicians" who want to isolate and learn well-defined procedures, and "connectors" motivated by a need to understand basic principles (terminology of Frid, 1992). Even using inexpensive methods of assessment, we have to find ways to create more students who are "connectors" if we are to have any hope of leaving them with something of value that endures for more than a few days after the final examination.

Part IV

Assessing Understanding of Probability

Chapter 17

Dimensions in the Assessment of Students' Understanding and Application of Chance

Kathleen E. Metz

Purpose

Randomness and chance variation are key ideas that can function as goals in young students' understanding and application of chance. In this chapter I examine how these key ideas involve construction of new concepts, as well as beliefs about the place of chance in the world. These ideas are considered from the perspective of the mathematics or statistics classroom culture; i.e., how the classroom culture reflects and fosters beliefs about the place of uncertainty and chance in the world.

IDENTIFICATION OF KEY IDEAS

The National Council of Teachers of Mathematics' *Assessment Standards for School Mathematics* (1995) asserts that current classroom assessment practices frequently tap the "uninteresting and superficial." According to this document, we need to fundamentally reframe our assessment tools, so that they reveal students' knowledge of core ideas and how students apply these core ideas to new and familiar problems. This criticism of contemporary assessment practices parallels NCTM's (1989) concerns, delineated in the *Curriculum and Evaluation Standards for School Mathematics*, that contemporary instructional practices pay too much attention to algorithms and too little attention to the concepts and big ideas underlying the algorithms.

Instruction and assessment in statistics and probability have frequently constituted an extreme example of a focus on procedures to the neglect of underlying concepts and big ideas. In this vein, Stodolsky argues (1985) that even university students are frequently taught statistics with the assumption that experts will tell these students what to do if and when they subsequently need statistics for their own purposes; thus conceptual understanding appears neither expected nor necessary. One important step in the strengthening of instruction and assessment in the domain of statistics and probability involves identification of concepts and big ideas as a key frame in the analysis of the conceptual underpinnings of the curricula and the adequacy of students' evolving knowledge in the domain.

The national curricular reform documents of Australia, Great Britain and the United States each recommend that students in the first level of schooling reflect on ideas of chance and probability, within the context of interpreting data they have collected themselves (Romberg, Allison, Clarke, Clarke & Spence, 1991). Analysis of the psychological and mathematical literature concerning chance reveals the deep complexity and multifaceted aspects of these constructs and their application. For the purposes of more specifically identifying the big ideas in this sphere that are within reach of young students, we find extremely useful the thinking of statistician and educator, David Moore (1990). He points us to two constructs, randomness and chance variation, that he argues can be fostered through appropriate childhood experiences.

Randomness

Both the cognitive and mathematical literature manifest considerable disagreement on the meaning and entailment of randomness. For the purposes of assessing (young) students' curriculum and emergent learning, we use a basic definition offered by Moore. Moore explains:

> Phenomena having uncertain individual outcomes but a regular pattern of outcomes in many repetitions are called *random*. "Random" is not a synonym for "haphazard," but a description of a kind of order different from the deterministic.

Note that the construct of randomness and its attribution to a particular phenomena assume two aspects: the uncertainty and unpredictability of the single event, in conjunction with the patterns that emerge across a large number of repetitions of the event. Thus, although certainty of prediction is unattainable for individual random phenomena, at least prediction becomes possible over the long haul.

Chance variation

A major issue that emerges in any context of data interpretation is the question of whether the variations in the data have some kind of causal base or are simply due to chance. This issue can emerge as young students consider what the data they have collected do and do not mean, as well as in the more sophisticated interpretations of older students. Adequate interpretation of data demands an understanding of the idea that variations can be due to either chance and/or some form of underlying causality. Moore contends that it is crucial for students to grasp the idea that "chance variation, rather than deterministic causation, explains many aspects of the world" (Moore, 1990, p. 99). This key idea is fundamental to students' negotiation of data-based curricula, as well as their adequate interpretation and prediction of patterns outside of school — from accidents to sequences of shots on the basketball court.

THE CHALLENGE OF ASSESSING STUDENTS' UNDERSTANDING AND APPLICATION OF CHANCE

Assessment of students' understandings poses considerable challenges for teachers. Most simply, many kinds of tasks used in the instruction and assessment of statistics and probability may reflect little about students' underlying conceptual knowledge, but rather their ability to apply the right algorithm. For instance, consider an exemplar of a statistical activity for grades 3 - 4 given in NCTM's *Curriculum and Evaluation Standards*:

> "[Given a spinner divided evenly into four sectors, with three sectors colored red and one sector colored blue] "Is red or blue more likely? How likely is yellow? How likely is

getting either red or blue? If we spin twelve times, how many blues might we expect to get?" (NCTM, 1989, p.56)

In this kind of problem, right answers could easily hide students' failure to grasp how chance enters into probabilistic situations. For example, in relation to the last question posed, the answer of "9 reds" may well reflect analysis of the ratio of colors in the spinner. However, this response does not reveal whether the student has grasped the idea that although 9 is indeed the best prediction, this particular outcome is uncertain. Without a keen attention to the reasoning underlying students' predictions— as revealed in class discussions or written work— the teacher would have little information about students' understanding of chance.

Above and beyond the significant conceptual challenges of this domain, appropriate application of chance also involves beliefs. Whether or not students will think to interpret a situation in terms of chance will be influenced by their individual beliefs about the place of chance and uncertainty in the world (Nisbett, Krantz, Jepson, & Kunda, 1983). Furthermore, the individual's beliefs about appropriate data interpretation and how chance may enter in are also influenced by the classroom culture and the beliefs and values it reflects and encourages.

In an effort to address these complexities, this chapter analyzes assessment of the understanding and application of chance along three dimensions: *(a)* cognitive: the student's conceptual constructions; *(b)* epistemological: the student's beliefs about the place of chance and uncertainty in the world; and *(c)* cultural: the culture of the mathematics classroom and learning environment vis à vis beliefs about how chance and uncertainty fit in and related dispositions of interpretation and inference. Each dimension is examined through consideration of its relevance, analysis and illustration of relatively naive versus more adequate stances, and its implications for assessment. Illustrative examples are based on the authors' research and work with children in the early grades, but the points raised are general and for the most part apply to older students as well.

THE COGNITIVE DIMENSION: CONCEPTUAL CONSTRUCTION

A more conceptually-oriented approach to students' study of chance and their interpretation of variability in data demands assessment of students' conceptualizations of the core idea of randomness and the alternative interpretations that students may evoke instead. The manner of teaching now encouraged by the mathematics reform movement — including fostering of mathematical discourse among students and critical analysis of the meaning of data they collect — manifests a range of rich indicators of students' thinking that can reflect the distinctions in students' understanding of chance described below.

The challenge of understanding randomness

As noted above, randomness involves both the uncertainty and unpredictability of a single event, with an understanding of the patterns that can emerge over a long sequence of repetitions of the event. The challenge of integrating uncertainty and pattern aspects into the randomness construct constitutes a primary issue in cognitive development and a primary source of errors or "bugs" among older students and adults.

The cognitive literature has documented young students' attribution of uncertainty and unpredictability to what are actually random phenomena, without a corresponding sense of the patterns involved. Conversely, it has also documented young students' attribution of deterministic patterns to random phenomena, without any sense of the corresponding element of unpredictability and uncertainty. While a *patterns-without-uncertainty* interpretation exaggerates the information given, an interpretation of *uncertainty-without-patterns* underestimates the information given. One fundamental issue to consider in the assessment of students' grasp of randomness is the status of the construction; i.e., evidence of the randomness construct in integrated form versus evidence of an interpretation limited to patterns or uncertainty.

Extending the construct of randomness across probability-types

Probability is not a singular construct, as evident explicitly in the perspective of the expert and implicitly in the young student's various ways of deriving and justifying probabilities. The focus below is on the two types of probability that most frequently arise in the current approaches to teaching students chance and probability: *classicist probabilities* and *frequentist probabilities*. We have no reason to believe that the different ways of deriving and conceptualizing probability pose equivalent challenges to students. Furthermore, we have no basis to assume that students' understanding of one form of probability entails any understanding of other probability-types.

Classicist probabilities

The use of chance-generating devices, such as die, coin tossing, or spinners constitutes a common instructional technique for introducing the idea of randomness. Classicist probabilities denote the derivation of probabilities from an analysis of the symmetries in chance-generating devices. For example, as a coin has two sides, it has a probability of .5 of landing on tails and .5 of landing on heads. Indeed, such games, analyzed in terms of their fairness, appear early in many probability curriculum materials for children and adolescents. This approach enables relatively efficient data collection and relatively straightforward derivation of expected probabilities, in the form of classicist probabilities.

Frequentist probabilities

This type of probability denotes the frequency of a given outcome over an infinite number of repetitions of the event. It can be approximated by conducting experiments that involve repeating events many times, and noting the relative frequencies that emerge over many trials. Interpretation of these data and derivation of probabilities would not be amenable to a classicist perspective, but will instead draw on children's intuitions of frequentist probabilities. (As noted above, US, English and Australian mathematics education documents recommend that children conduct simple investigations about themselves and the world around them.)

As frequentist probabilities presume some understanding of the notion of infinity, this probability-form may be particularly challenging to young children (Hawkins & Kapadia, 1984). There is evidence that aspects of the construct of infinity are difficult for school-age children (Piaget, 1987; Nunez, 1993). Nevertheless, children can grasp the precursor construct of probabilities in terms of the distributions that emerge across many repetitions.

ANALYSIS OF STUDENTS' COGNITIVE CONSTRUCTIONS

The following sections focus on the analysis of student's cognitive constructions, from the perspective of the status of the randomness construct and probability-type. First, examples of patterns without uncertainty are considered, followed by uncertainty without patterns, and finally uncertainty with patterns combined in the construct of randomness. Each of these cognitive constructions is illustrated in the form of classicist and frequentist probabilities.

Patterns without uncertainty

Students can reason in terms of patterns, but fail to conceptualize the chance involved in these patterns. Hence they exaggerate the information given. The interpretation of patterns without uncertainty in classicist form is illustrated through a child's analysis of the effect of different spinners on game outcomes. In this example, the child derives absolutist predictions of outcomes from the symmetry of the spinner-device:

> The teacher shows Mary, a third grader, two spinners. One of the spinners is evenly divided into two sectors, one red and one yellow. The other spinner is 3/4 red and 1/4 yellow. The teacher explains that a player gets to advance up the game-board when the spinner lands on their color during their turn. She then asks Mary, "Do you think it will make any difference which spinner we play with?" Mary argues that it will make a difference, because when you play with the spinner with the even division of red and yellow, "Sometimes you turn it, it'll stop on the red and sometimes you turn it it'll stop on the yellow."

> She denies this uncertainty applies to the other spinner with the asymmetrical color division. When the teacher asks "When you turn this [3/4 red spinner; 1/4 yellow] does it sometimes stop on the yellow and sometimes stop of the red?" She shakes her head. "No." When the teacher asks her to predict where the player with the yellow color will be at the end of the game, Mary predicts the board starting position.

This excerpt reflects patterns without uncertainty, in that the student infers that spatial predominance determines each outcome (as indicated by an analysis of the symmetries in the outcome-generating device). Less primitive forms of patterns without uncertainty involve the inference that the outcome associated with spatial predominance will always win overall, but the outcome associated with a subordinate spatial value will occur periodically and any particular spin is uncertain.

In terms of the challenge of assessment, note that prediction of who would be most likely to win the game would not have differentiated Mary's buggy construct from the adequate randomness construct: both interpretations would have predicted the predominant color as the winner. The inadequacy of Mary's conceptualization of chance is manifested in her predictions about the placement of the losing player, and more generally, the certainty which she attributes to all of her projections.

Patterns without uncertainty as a precursor to frequentist probabilities is illustrated through excerpts of a student's approach to a sampling problem. In this situation, the student views the distribution of elements in the small sample drawn as an accurate reflection of the urn contents. The missing aspect of uncertainty takes the form of the student's failure to recognize how chance enters into the sample viewed. More generally, this manner of thinking reflects what Tversky and Kahneman (1971) have facetiously called the Law of Small Numbers:

> The teacher shows an urn, filled with a hidden collection of marbles to Janice, another third grader. She asks Janice if she can figure out what's in the urn without dumping them all out and without looking. Since Janice has no idea how to proceed, the teacher suggests she iteratively take one out, look at it, and replace it. The hidden collection actually has an even number of red and blue marbles. She supplies Janice with a collection of chips the same colors, to help her keep track of her marble draws.
>
> Janice makes 12 draws with replacement, each time adding another red or blue colored chip to her piles to keep track of her draws. She concludes, "I think it's mostly red, because there's seven of the red and 5 of the blue." When she checks her conclusion by peeking into the jar, she looks astonished at her discovery that there are not more reds. Janice explains that she had thought there were more reds, "Because there was more red [chips] than blues [chips] on the table". Upon probing, Janice denies there is any way she could have figured out there was the same number or any way she could have been more sure of her prediction.

In this case, the student's patterns without uncertainty interpretation is reflected by her confident interpretation of the meaning of small differences in the subsets of the sample, her bewilderment when she discovers her inference is wrong, and her failure to consider the potential of extending the action that the teacher viewed as the sampling procedure.

Conceptualizations of patterns without uncertainty are reflected in a number of buggy interpretations identified even at the adult level. The Gambler's Fallacy or negative recency affect denotes the expectation of local corrections to random fluctuations in a sequence; e.g. the assumption that the toss of two tails in a row will be followed by the toss of a head. Vallone and Tversky (1985) have identified a related bug, where the individual assumes random fluctuations in the data must be causal and proceeds to develop causal explanations.

Uncertainty and unpredictability without patterns

Whereas in patterns without uncertainty, students assume they can confidently predict what will happen, thereby exaggerating the information given, in *uncertainty without patterns* students conceptualize the situation as simply unpredictable, thus underestimating the information given. In the situation with the spinners, uncertainty without patterns is manifested by the belief that as long as both color outcomes are included in the spinner, the situation is impossible to predict:

> The teacher asks Karen, a kindergartner, if she thinks it'll make any difference which spinner they play with. Karen argues, "No," 'cause they both have yellow and both have red." The teacher suggests they play with 3/4 red, 1/4 yellow spinner and invites Karen to choose her color. She selects yellow for herself, the color of the bad odds, and has no predictions about who might win.
>
> After the child has lost the game, the teacher says, "I want you to look real carefully and see if you think it makes any difference which one you play with." Karen maintains her

position: "Nah, it wouldn't because they [the spinners] both have yellow and they both have red." "Do you think it makes any difference if there is more red on one than the other?" the teacher hints. "No", Karen argues, "Because they have yellow and they both have red."

In this example, Karen's buggy conceptualization of uncertainty without patterns is reflected in her choice of the spinner with the bad odds for her color, her absence of predictions for the game outcome and, more generally, her conviction that it is irrelevant which spinner they play with, as long as the spinner has both players' colors.

The failure to anticipate the differential effect of asymmetries in the chance-generating device is relatively primitive. Most kindergartners do anticipate differential effect, although typically without a corresponding appreciation of the chance involved. However, uncertainty without patterns is manifested by much older students and even well-educated adults in connection with frequentist probabilities.

Both children and adults reflect this manner of thinking in the urn sampling problem. In contrast to the spinner situation, many sampling problem-types involve a sampling space initially in the form of an unknown. Children and adults alike can focus on the collection in the urn as an unknown and fail to appreciate how a succession of draws can inform them about its contents:

Third grader, Nathan, continues to be skeptical about the feasibility of the urn sampling task, even after the teacher suggests the procedure of drawing and replacing a marble at a time: "I don't think that'll work because I might like have taken the same marble out a couple times." "Do you think if you did it a *whole* bunch of times you could figure it out?", the teacher suggests. "No," Nathan insists. "It's impossible."

In his acute awareness of the uncertainty in the situation, Nathan has failed to realize that patterns can emerge across extended repetitions of the random event and that these patterns do give us some information, although imperfect, about the population from which they were drawn.

The construct of randomness: learning and assessment

Application of the construct of randomness demands an integration of the uncertainty and unpredictability of a single event, with an understanding of the patterns that can emerge across repetitions of events. The cases cited here were from students who had manifested considerable learning on task. The first example is Karen, who had previously argued that, as long as both players' colors were on the spinner, the game was unpredictable. Consider her reasoning four games later:

In conjunction with the two-colored 3:1 red/yellow spinner, Karen chooses the red chip for herself, relegating the yellow chip to the teacher. Karen explains she wants to play with the red chip, "Cause the red has more than the yellow. Because if you play with the red, you have a very good chance of winning."

Karen's grasp of randomness is reflected in her choice of the chip with the better odds, in conjunction with her explanation of its affect: her probable but uncertain win. Her derivation of differential probabilities is based on an analysis of spatial relations in the chance-generating device.

Similarly, recall that Nathan had previously contended that prediction of the urn contents was impossible simply on the basis of single draws. With scaffolding, Nathan begins to integrate notions of uncertainty and unpredictability with the idea that patterns that can emerge across large numbers of draws. He remains convinced that, no matter how long one continues sampling, you can never know for sure:

> Although skeptical of its utility, Nathan agrees to try the procedure of one-by-one sampling with replacement that the teacher suggests. After drawing 7 marbles, including 6 reds (which Nathan considers orange) and one yellow, he comments, "I think they're more of the oranges now because I keep on taking out oranges." "Are you quite sure?" probes the teacher. "Not really," he responds.
>
> When the teacher asks Nathan if it would make any difference if he continued the draw and replacement procedure, he immediately picks up on the idea, exclaiming, "I think it would give me a better idea!". After 8 more draws, consisting of 7 oranges and one yellow, Nathan concludes, "I think it's more orange." "Are you quite sure?" questions the teacher. "Uhhuh," he affirms. "Any way you could be very sure or is that impossible?" she probes. "That's impossible!" exclaims Nathan.

Note that Nathan's prediction ("It's more orange") could stem from either a deterministic interpretation of patterns or an integration of chance and patterns. It is his qualifications concerning his confidence in this prediction, together with his assertions that certainty is unattainable, that indicate the integration of chance and patterns in the randomness construct.

When teachers focus classroom discourse on such issues as the meaning of patterns and variability in the data, and the bounds of predictability and control, they can simultaneously support the learning and assessment process. Teachers can use these discussions to analyze the students' grasp of randomness and chance variation. These constructs are fundamental in their learning to successfully cope with uncertainty, in experiences beyond the school walls as well as their formal study of statistics.

THE EPISTEMOLOGICAL DIMENSION: BELIEFS ABOUT CHANCE AND RANDOMNESS

The differentiation of deterministic from nondeterministic events challenges us all; children and adults, novices and experts. Indeed, the demarcation of the causal from the stochastic stands at the forefront of many fields, including genetics, physics and ecology. Analysis of intellectual history (Gigerenzer, Swijtink, Daston, Beatty, & Kruger, 1989; Hacking, 1990) and contemporary bugs in adults' thinking (Kahneman, 1991; Kahneman, Slovic & Tversky, 1982) reveal a general tendency to err toward the side of attributing too much to deterministic causality, with a corresponding failure to recognize the extent to which chance operates in what one experiences of the world.

If students are to appropriately use the interpretative frame of *chance variation*, they need to both grasp the idea of randomness *and* have a view of the world with a place for chance events. In other words, a student may have grasped the idea of randomness, but unless she believes that some phenomena are indeed due to chance she will not apply the construct. This epistemological dimension of chance is relevant to teachers, in that variations in epistemological stance of this kind can affect students' ability to grasp these big ideas, as well as their propensity to apply chance interpretative schemas. This chapter uses the term *epistemological set* to denote an individual's predilections to interpret the world in relatively deterministic or stochastic terms, based on their beliefs about the place of chance in the world.

Epistemological set does not imply that individuals do not have both chance and deterministic interpretations within their cognitive repertoire, but that they have a tendency to interpret phenomena toward one or the other end of the stochastic / deterministic continuum. As Fischbein argued, "What we are really and specifically concerned with here is the *orientation of knowledge*, a factor of cognitive 'set' or a kind of 'mentality' on the intellectual level." (Fischbein, Pampu & Minzat, 1975, p. 170).

This chapter examines students' epistemological set, as a second dimension in teachers' assessment of students' understanding and application of chance. The sections below consider examples from students with epistemological sets from both ends of the continuum and the nature of the indicators that differentiate them.

Students with relatively deterministic epistemological sets

Data interpretation of two different students are used below as illustrations of relatively deterministic epistemological sets. The first case is Christy, a six year-old kindergartner. Her data interpretations are reported from two tasks. One task, adapted from Piaget and Inhelder (1975) involves predictions and interpretations of the mixing of marbles, initially arranged by color, in a box that can tilt up and back. The other task is the urn sampling problem described above.

The teacher shows Christy a shallow box affixed on an axis. Along the lowered edge of the box are 12 marbles, including 6 black ones on the right and 6 silver ones on the left. The teacher asks whether or not the marbles will return to their original positions after the box is tilted back and forth. After initial predictions about movements of the marbles upon rotation of the box up and back, Christy takes the box from the teacher, telling her she wants to figure out what's happening. Christy exclaims, "They're trying to get back to their places! These [black marbles] want to get over here and these [the silver marbles] want to be right here! " The teacher probes, "Why do they want to be in those places?" Christy explains, "Because that's the way God made them!"

Christy's assumption that the marbles are trying to return to their places reflects a view of the world as ordered. It doesn't occur to her that the marble movements may be affected by chance factors.

In the urn sampling problem, Christy's interpretations again manifest a relatively deterministic epistemological set. Here her deterministic epistemological set is manifested first through the topology of the bottom of the urn that she argues determines which marble is drawn next; and subsequently through the potential control one has over marble color drawn.

After just 4 draws, Christy is confident that there are more greens than reds. "Are you quite sure?" probes the teacher. "Yah!" she immediately responds. Christy then inspects the urn contents to discover the same number of reds and greens. The teacher asks her, "Why did you think there were mostly green when actually there were the same number?"

> In response, Christy appears to assume a kind of pathway in which the marbles are arranged by color and thus proceed one color before the other: "Because it was like this. Every time I would drop it in front, so I would keep on getting green until I got a red. Maybe when I picked up, I always picked up green, green, green green, until the reds finally came ahead."
> Finally, Christy suggests that she could have gotten it right by strategic marbledraws: "They could maybe go different by different. First go red, green, red, green." The teacher questions, "But how would you be able to do that?" "Maybe if I pulled in different directions," reasons Christy.

In her pathway model of marble selection, the selection process would be largely determined. Of course, this model totally ignores the random nature of the arrangement in the urn. Her idea that she should have been able to draw the marbles in an alternating order, such that the equality of the different colors in the urn was transparent, also fails to appreciate the essential random nature of the draw.

The second example of a relatively deterministic epistemological set is drawn from one child's thinking within a class-based investigation, suggested in the Used Numbers book, *Statistics: Middles, Means and Inbetweens* (Friel, Mokros, & Russell, 1992). A fifth grade class has been collecting, representing, and interpreting data concerning the number of raisins in a box. The teacher has asked the students to formulate their best guess concerning the number of raisins in a box, given data about the contents of 28 other boxes they have already opened.

> Building on a representation of the data invented by a small group of students, the teacher has organized the class data into a line plot. She shows the students the line plot and then challenges them to figure out what would be the best prediction for the number of raisins in an unopened box. Sue's strong prediction is 72. She argues, "There is no 72 [in the distribution to date]! And it's really close to the average [the mean has been calculated as 73]. You kind of wonder why there isn't 72. You have to wonder why there isn't a 72. I think the next box will have 72."

It doesn't occur to Sue that the absence of any 72 data point could be due to chance. In assuming that the dip in the curve will be filled by the next data point, Sue's reasoning reflects a spurious determinism. She fails to grasp the fact the each raisin box sample is an independent event, not affected by the particular boxes that happen to have been counted previously. Her thinking is closely related to the common adult bug called negative recency effect or Gambler's Fallacy, in which one anticipates local corrections to random fluctuations. Again, the student's answer alone would not have revealed her deterministic set. This orientation becomes evident in her explanation of why she infers 72 would be the best answer.

Students with epistemological sets emphasizing chance and uncertainty

Two illustrations of epistemological sets emphasizing chance and uncertainty are considered below. The first case is Nathan, whose acute sense of the unknown and the uncertain was noted above in connection with the urn sampling problem. Nathan's thinking in relation to the marble tilt box (see Christy's case in the last section) and the spinner reflect a similar emphasis of chance and uncertainty, *albeit* in different forms:

In the marble tilt box problem, Nathan assumes some kind of unpredictable change in the marble arrangement. He explains, " The marbles will lose direction. They will loose control." In response to the teacher's query of whether or not the marbles will return to their original positions, Nathan will not commit himself beyond "Could be." When asked how long this might take, he sticks to "I don't know."

In the spinners task, Nathan is guarded about all his predictions. For example, when analyzing the affect of playing with the 3:1 red/yellow spinner, Nathan notes "I *think* with that color [red] I'd win, because it's [the spinner] more of my color." Even when he is way ahead of the teacher, also burdened with bad odds, he predicts he'll win, but also notes she still could win "If you keep on hitting on the yellow."

In my experience of interacting with Nathan across six different activities, he was consistently prone to interpret data or predict outcomes in non-deterministic terms. Chance and uncertainty appear to enter into Nathan's view of the world, from both his assumptions of how the world works as well as intense awareness of his own ignorance.

The second case is John, another third grade boy who also appears to view the world in terms emphasizing chance and uncertainty. John's perspective is briefly described in relation to the marble tilt box problem, analysis of the spinners, and the urn sampling problem.

John is reluctant to commit himself on a prediction concerning what the marbles might look like after a rotation up and back, simply commenting "It might not [come back to the black marbles on one side and silver marbles on the other]. The teacher's query, "Do you think eventually the black ones will be on the other side from where they started out?", leads to a similar noncommittal response: "It might happen some time."

In the context of the spinner games, John spontaneously tells the teacher, playing with the yellow chip in conjunction with the 3:1 red/yellow spinner, " You have a chance. . . . if you get lucky."

John is concerned about the feasibility of prediction of the urn contents, "Because every time you put it [the marble drawn] back, you might get the same color as before, so you can't be sure." The teacher suggests, "If you kept on doing that for a long time, would you be pretty sure [if they're mostly one color or the same number of each]?" John remains tentative, telling the teacher, "You never know."

John seems attuned to the uncertainty of events and, conversely, reluctant to make firm predictions on the basis of anticipated patterns. His epistemological set, emphasizing chance and uncertainty, is manifested in his reluctance to make predictions, his acknowledgment of the possibility of multiple outcomes, and his strong sense of the unknown and unknowable.

Implications for Assessment

Epistemological set — i.e. the students' beliefs about where and to what extent chance enters into the world — are relevant to the classroom teacher who teaches statistics, in that it will strongly affect the students' propensity to consider a stochastic interpretation. Epistemological set can be reflected in the assumptions students make about causality and chance and their respective spheres of application. Differences will be manifested in discrepant ways in which students interpret and explain the same data set: seeing determinism in variability versus seeing probabilistic patterns or uninterpretable uncertainty. These distinctions in epistemological set can be reflected in whole class or small group discussions about appropriate data interpretation, where (as in the example of Sue's class noted above) students not only give their answers but discuss why they consider their interpretation the most reasonable. More generally, students' epistemological set is revealed in their propensity to assume deterministic explanations versus their willingness to seriously consider the possibility that the outcomes may be due to chance.

THE CULTURAL DIMENSION: HOW THE CLASSROOM ENVIRONMENT TREATS CHANCE, UNCERTAINTY, AND BELIEFS

I had to rethink my interpretation of Christy's remark that God made the marbles move the way they had after her principal asked me, "You don't *really* believe that anything is due to chance, do you? We teach our children that God is behind all things." To adequately understand students' cognitive constructions and beliefs, we need to consider the culture in which students participate. The extent to which individuals assume that deterministic causality underlies variability, as opposed to the possibility of random variation, depends in part upon the orientation of their society at large and school culture. According to Nisbett, Krantz, Jepson and Kunda (1983), the culture influences the kinds of situations in which individuals are likely to consider a stochastic interpretation.

We can conceptualize students' learning statistics and probability as in part a process of enculturation. Indeed, Lauren Resnick has argued that becoming a mathematical problem-solver is in large part a process of enculturation, "as much acquiring the habits and dispositions of interpretation and sense-making as of acquiring any particular set of skills, strategies, or knowledge" (Resnick, 1988, p. 58). One subtle yet important parameter of assessment in this domain concerns the extent to which the culture of the math classroom, in the activities it structures and the interpretations it values, embodies a deterministic versus nondeterministic view of the world. We can conceptualize this dimension of assessment as a kind of epistemological set at the level of the social group within the social institution of the school. Below we consider assessment from the cultural perspective of messages about the place of chance and determinism, implicit in the values and habits of the classroom learning environment.

The task of assessing the culture of the mathematics classroom in terms of epistemological set involves a shift in emphasis from learner to learning environment. Indicators of classroom epistemological set include the choice of subject matter, the structuration of problems, appropriate means of data interpretation, sufficient evidence to assume causality, and the aesthetics of what constitutes a good solution or explanation.

Classrooms that support a deterministic view of the world are easy to find. A very prevalent method of teaching students mathematics has been and continues to be teacher telling, followed by text-based students' practice (Goodlad, 1984; Stodolsky, 1985). Within this approach, there are right answers and wrong answers, with little ambiguity about what algorithms should be applied. Indeed, the school culture has been identified as, in general, a strong influence in the direction of a deterministic epistemological set. Moore (1990) points to the choice of curricular emphases in the initial years of children's formal schooling as a factor in the building of a deterministic world view, particularly in the realm of mathematics. Moore writes:

> Children who begin their education with spelling and multiplication expect the world to be deterministic, they learn quickly to expect one answer to be right and the others wrong, at least when the answers take numerical form. Variation is unexpected and uncomfortable (Moore, ibid., p. 135).

In addition to the fact that the basic skills curriculum tends to be strictly deterministic, where ambiguity does come in (e.g., the best representation to use in solving a word problem), the authority in the guise of teacher or text provides the single correct form.

More specifically, Fischbein (1975) argues that schooling tends to have detrimental effect on children's understanding of chance. Indeed, this researcher points to evidence in his work and in unexplicated findings in the work of others of a regression in children's grasp of randomness after the point of entry into formal schooling, a regression which he attributes to the influence of the school culture. Fischbein asserts:

> It is generally true that contemporary education oversimplifies the rational, scientific interpretation of phenomena by representing it as a pursuit of *univocal*, causal or logical dependencies. Whatever does not conform to strict determinism, whatever is associated with uncertainty, surprise, or randomness is seen as being outside of the possibility of a consistent, rational, scientific explanation... . The intuitions of chance and probability are influenced, and ultimately deformed, by this excessive tendency toward univocal prediction" (Fischbein, 1975; pp. 124-125).

Fischbein contends that the teaching process tends to either reject chance events as unexplicable or to impose deterministic relations where they do not exist.

From the perspective of mathematicians as well as the NCTM reform documents, conceptualizations of mathematics education should emphasize processes such as exploring patterns, inference from data, formulating conjectures, etc. It is within this conceptualization of mathematics education that chance and ambiguity enter in. Within this vein, Resnick (1988) has argued that we should not teach mathematics as a "well-structured discipline" where students practice the rules of the domain, but rather as an "ill-structured discipline" where students construct multiple strategies and viewpoints and corresponding justifications of their sensibility.

In illustration of how a classroom learning environment can reflect this kind of epistemological perspective, a collaborative activity within a primary level classroom is briefly described. The investigation was based on a suggestion in the Used Numbers' book, *Counting Ourselves and Our Family* (Stone & Russell, 1992).

A teacher challenges her combined first-second grade class to try to figure out how many children there are in a particular second grade classroom, when all the children are out of the room (and cannot be directly counted). The need for this information is framed in terms of how many cookies they should make to have enough for themselves and the other class. The children begin exploring the problem by reflecting on possible indicators of class size. For example, after one child suggests they could count pencil boxes, an informant tells her classmates that this won't work because some children share pencil boxes in this classroom. Several children think a chair count would be a good idea.

The following day, equipped with a clipboard for each pair of investigators, the children enter the empty second grade classroom. Most pairs count items that could well correspond with the number of children in the classroom, including chairs, number of children indicated on a bar graph of birthdays by month, children's writing folders, and children's names in a poster. One pair of children count the number of Kacheena Dolls on display. They duly record the total (9), apparently oblivious of the fact that this number is a fraction of the size of any class in the school. Another pair counts a total of 13 backpacks, but immediately questions the reliability of the count as an indicator of class size, on the grounds that "lots of kids don't bring backpacks to school".

Upon return to their own classroom, the children and teacher make an ordered list of the counts they have found. The children record their counts on small squares of sticky paper (5 cm X 5 cm). The teacher constructs a bar graph, to which each pair affix their data point(s), explaining what they have counted to reach this sum. On the basis of the bar graph, the children infer there are 28 children in the other room, because 28 has the most counts.

When the children deliver the cookies, they discover that their inference was exact: the class indeed has 28 children. The teacher and a minority of the children display genuine excitement at their collaborative detective (and mathematical) work. Other children express frustration that they have failed, in that their count, the data point they contributed to the bar graph, was wrong.

Conflicting values are reflected between the teacher and the activity she has guided and many of the children new to her classroom. The teacher's task structuration and the modes of interpretation she encourages reflect an assumption that although a single count may or may not reflect the true number of children in the room, analysis of data collected across many counts of different indicators may well accurately reflect the correct number or approximate it. Furthermore, she does not lead her children to try to account for every blip in the distribution, but rather to examine the distribution for trends at the same time as they consider the issue of validity underlying outliers in the distribution. This orientation contrasts with some children's assumptions about mathematical activity; namely that, by correct application of the correct [counting] algorithm, they should have been able to get the right answer themselves. Indeed, if mathematics is simply application of the correct rule, their assumption is valid. However, in this classroom engaging in mathematics demands much more than competent algorithm-application.

Implications for assessment

Above and beyond our assessments of the learner, we need to examine how the learning environment supports or subverts the big ideas and interpretative schemas of the chance curriculum. The culture of the classroom reflects a kind of collective epistemological set, with varying degrees of convergence or conflict. Assumptions about the degree of determinism underlying phenomena or conversely the place of chance and uncertainty in the world can be reflected in the model of mathematics implicit in the curricula, the structure of problems presented to the children, practices of interpretation, and the goal-structure of classroom discourse.

IMPLICATIONS

In summary, this chapter examined assessment of students' emergent understanding and application of probability from three perspectives: (a) two key concepts underlying knowledge of the domain; (b) how beliefs come into play in whether or not students think to apply these ideas in their attempts to make sense of patterns, and (c) how the classroom culture can support or subvert students' grasp and utilization of the big ideas underlying the statistics curriculum. The chapter also examined how teachers can assess students and classroom learning environments along these dimensions. In short, unlike most other subject areas, the teaching and learning of statistics is remarkably complex, involving both new and difficult concepts and beliefs systems that are resistant to change. Teaching of statistics can be empowered by seriously considering these various perspectives and the dynamics of their interaction.

Challenge of grasping the key constructs of randomness and chance variation. In the spirit of NCTM's recommendations that we approach the teaching of mathematics and statistics from a more conceptual perspective, identification of key concepts within the various curricular strands becomes crucial. Randomness and chance variation constitute such fundamental concepts in the context of the statistics and probability strand. The research literature reveals that these rich, multifaceted concepts can be a fruitful focus of reflection from kindergarten through high school.

Multiple dimensions of assessment: cognitive, epistemological and cultural. The goal of supporting students' appropriate interpretation of stochastic phenomena— or more specifically, their appropriate attribution of randomness and chance variation— is actually quite complex. First of all, a cognitive dimension comes into play, due to the fact that attainment of this goal presumes cognitive constructions of remarkably subtle concepts. Second, an epistemological dimension comes into play as, to a greater extent than most domains in mathematics, appropriate interpretation of stochastic phenomena also involves the students' beliefs, beliefs about the extent to which chance explains phenomena in the world. Finally, a cultural dimension comes in, as the students' beliefs about the place of chance and, more generally, stochastic versus deterministic explanations, will presumably be influenced by the values, dispositions and habits of the school and classroom culture. All three of these dimensions directly affect the success of our statistics instruction.

Assessment on the cognitive dimension. Obviously, the classroom teacher seldom has the luxury of working one-on-one with individual students to diagnose their understandings. Fortunately, small group and whole group discussions can provide a rich window onto their various understandings and beliefs. For example, returning to NCTM's question about "How many blues might we expect to get?" (NCTM, 1989, p.56), in 12 spins with a evenly-divided four sector spinner, discussions about whether or not they can predict what will happen exactly and why or why not, the source of uncertainty and variability, followed by their collecting of spinner outcome data and reflecting on how and why these data accord with and depart from their predictions, can provide a rich learning and assessment activity. In short, these crucial dimensions of students' conceptual understandings can be reflected in mathematics classes that emphasize students' explication and collaborative analysis of their emergent ideas, conjectures, reasoning processes and argumentation.

Assessment on the epistemological dimension. If we are to foster appropriate utilization of a stochastic interpretative frame, we need to also assess students' beliefs about where and how chance may enter in. Above and beyond having constructed the concept of randomness and chance variation, whether or not students will think to interpret a situation in terms of chance depends in large part on their beliefs about the place of chance and random variation in the world. For example, beliefs about the place of chance in the world may well enter into students' propensity or willingness to qualify their calculation of "9 blues" with the idea that chance is also involved here — many times the exact answer will not be 9. Furthermore, both conceptual constructs and belief systems may well affect their ability to grasp such subtle ideas as expected value — in this case, that 3/4 is a ratio of the expected relative distributions over an infinite number of repetitions of the event, but a ratio that is only approximated across many repetitions. In short, students' beliefs about how chance fits into the world is relevant to teachers in that students' varying perspectives affect their openness to interpretative principles based on chance, as well as their inclination to use them.

Assessment of the culture of the mathematics classroom vis a vis ideas of chance. The culture of the mathematics classroom is an influential force in what students do and do not learn about ideas of chance and the place of chance in the world. In addition to assessing the students' understanding and beliefs concerning chance, teachers need to consider how the practices and values of the mathematics classroom support or hinder the ideas about chance and its application that they are intending to teach. Analysis of this implicit level of the curriculum is particularly important given the growing recognition of the power of these hidden messages and their cumulative effect on students' thinking and beliefs.

Chapter 18

Combinatorial Reasoning and its Assessment

Carmen Batanero, Juan D. Godino and Virginia Navarro-Pelayo

Purpose

In this chapter we provide some answers to the following questions:

- What is combinatorics and what role does it play in teaching and learning probability?
- What components of combinatorial reasoning should we develop and assess in our students?
- Are there any task variables that influence students' reasoning and provoke mistakes when solving combinatorial problems?
- What are the most common difficulties in the problem-solving process? How should we consider these variables in the teaching and assessment of the subject?

We illustrate these points by presenting some examples and test items taken from different research work about combinatorial reasoning and samples of students' responses to these tasks.

WHAT IS COMBINATORICS?

The scope of combinatorics is much wider than simply solving permutation, arrangement, and combination problems. In his "Art Conjectanding," Bernoulli described combinatorics as the art of enumerating all the possible ways in which a given number of objects may be mixed and combined so as to be sure of not missing any possible result. According to Hart (1992), combinatorics is the mathematics of counting. It is concerned with problems that involve a finite number of elements (discrete sets), with which we perform different operations. Some of these operations only modify the set structure (e.g., a permutation of its elements) while others change the set composition (taking a sample). We are usually interested in a *combinatorial configuration* or composition of the result of some of these operations, and we attempt to answer the following questions:

- Does a specific combinatorial configuration exist?
- How many combinatorial configurations are there in a given class?
- Is there an optimum solution to a (discrete) problem?

These questions correspond to three different categories of combinatorial problems: *existence problems* deal with whether a given problem has a solution or not; *counting problems* investigate how many solutions may exist for problems with known solutions; *optimization problems* focus on finding a best solution for a particular problem. Considering Bernoulli's description, we must add *enumeration problems* that correspond to the question of whether we can produce a procedure for systematically listing all the solutions for a given problem.

The teaching of combinatorics is currently not considered necessary by many statistics teachers, probably because they restrict its meaning to counting problems and to combinatorial operation formulae. Nevertheless, in our teaching proposal (Batanero et al., 1994) we have distinguished the following components in the teaching and assessment of combinatorics.

Basic combinatorial concepts and models:

- Combinatorial operations: combinations, arrangements, permutations, concept, notation, formulae;
- Combinatorial models:
    Sampling model: population, sample, ordered/non-ordered sampling, replacement;
    Distribution model: correspondence, application;
    Partition model: sets, subsets, union.

Combinatorial procedures:

- Logical procedures: classification, systematic enumeration, inclusion/exclusion principle, recurrence;
- Graphical procedures: tree diagrams, graphs;
- Numerical procedures: addition, multiplication and division principles, combinatorial and factorial numbers, Pascal's triangle, difference equations;
- Tabular procedures: constructing a table, arrays;
- Algebraic procedures: generating functions.

Most of these contents are also linked to probability. Moreover, we can easily identify relevant statistics and probabilistic questions in each of the aforementioned combinatorial problem categories, as we can observe in the following classical situation of experimental design:

Example 1:

> Suppose you want to assess the effect of two different fertilizers on the improvement of tomato production. You have two types of tomato available and you would like to evaluate simultaneously the effect of low/high humidity degree on the production. Is it possible to design such an experiment using only 4 experimental plots?

In this situation, we can easily identify an example of a combinatorial *existence* problem. When trying to list all the different combinations of factors, we would be dealing with a combinatorial *enumeration* problem. If we ask for the total number of the three-factor combinations, we would be interested in a *counting problem,* for which the solution is 2x2x2=8, because in each factor (tomato plant, fertilizer, humidity) we have only two possible values. Finally, an *optimization* problem can be proposed when asking the number of different two-values factors that could be evaluated with only 4 experimental plots.

We may note in this example some features of many combinatorial problems that Kapur (1970) highlighted:

- Since it does not depend on calculus, it poses suitable problems for different grades.
- Usually very challenging problems can be discussed with students.
- It can be used to train students in enumeration, making conjectures, generalization, optimization, and systematic thinking.
- Many applications in different fields: chemistry, biology, physics, communications, number theory, etc., can be presented. Because of the interconnections between combinatorics and probability, which we shall discuss in the following section, this may serve to show students the applicability of probability in these different subjects.

THE ROLE OF COMBINATORICS IN TEACHING AND LEARNING PROBABILITY

Combinatorics is not simply a calculus tool for probability, but there is a close relationship between both topics, which is why Heitele (1975) included combinatorics in his list of ten fundamental stochastical ideas which should be present, explicitly or implicitly, in every teaching situation in the stochastic curriculum. This connection is noticeable in the main probabilistic topics of the primary and secondary mathematics curriculum, so an adequate level of combinatorial reasoning is linked to the attainment of the main curricular aims.

For grades 5-8, the Curriculum and Evaluation Standards of the National Council of Teachers of Mathematics (NCTM, 1989) recommended that the curriculum should allow students to:

- Model stochastical situations by devising and carrying out experiments or simulations to determine probabilities.

- Model stochastical situations by constructing a sample space to determine probabilities.

- Appreciate the power of using a probability model by comparing experimental results with mathematical expectations.

The concept of *random experiment* is the starting point in the study of probability at these levels. Two main aspects of a random experiment, according to Hawkins et al. (1992), are the clear formulation of the experiment and the identification of all its possible outcomes (the *sample space*). When describing simple experiments it is easy to list (enumerate) all the different outcomes of the sample space, but when we increase the number of trials, the enumeration processes can become very complex, and we may prefer to compute (count) the number of events. In both cases, we are dealing with a combinatorial problem.

The connection between combinatorics and the definition of probability is clear when using the "equally likely" approach to probability (Laplace's definition), which strongly relies on combinatorial techniques. In this approach, the probability of an event A is defined as the fraction $P(A)=N(A)/N$, where N is the total number of possible outcomes and $N(A)$ is the number of outcomes leading to the occurrence of A. According to Piaget and Inhelder (1951), if the subject does not possess combinatorial capacity, he or she is not able to use this idea of probability, except in cases of very elementary random experiments. The Curriculum and Evaluation Standards of the NCTM (1989) recommended the following (combinatorial) procedures for students to obtain these mathematically derived probabilities: building a table or tree diagram, making a list, and using simple counting procedures.

The *frequentist* approach to probability is based on experiments and simulations. Young children may use manipulative materials to determine experimental or empirical probability. By actually conducting an experiment several times, children determine the number of ways an event occurred, and by comparing those with the total number of experiments they obtain an experimental estimate of probability. Students in grades 9-12 can also do simulations, in which urn models and computer simulation are used to describe real experiments. Urn models are based on the idea of sampling, in which the definition of the combinatorial operations might be based. Moreover, students should develop a real comprehension of the power and limitations of simulation and experimentation only by comparing experimental results to mathematically derived probabilities, which frequently require combinatorial reasoning.

For grades 5-8, some examples of compound experiments, which are linked to combinatorial operations, are also suggested in the Standards of the NCTM (1989). The inventory of all the possible events in the sample space of a compound experiment requires a combinatorial constructive process from the elementary events in the single experiments. On the other hand, arrangements and combinations may be defined by means of compound experiments (ordered sampling with/without replacement or non-ordered sampling with/without replacement).

For grades 9-12, new probabilistic topics are included, in particular, discrete probability distributions. These distributions are expressed in many cases by means of combinations or permutations, as is the case of binomial or hypergeometric distributions.

To finish this section off, we recall that many probability misconceptions are related to the lack of combinatorial reasoning that often provokes the erroneous enumeration of the sample space in the problem. This suggests the need to help students develop their combinatorial capacity.

In the next section we shall analyze the main points on which the teaching and assessment should be focused. We use the following notation: $AR_{m,n}$ for the arrangements with repetition of m things, taken n at a time (respectively, Am,n for the arrangements without repetition, $CR_{m,n}$ for the combinations with repetition, and $C_{m,n}$ for the ordinary combinations).

STUDENTS' COMBINATORIAL REASONING AND ITS ASSESSMENT

Besides its importance in developing the idea of probability, combinatorial capacity is a fundamental component of formal thinking. According to Inhelder and Piaget (1955), combinatorial operations represent something more important than a mere branch of mathematics. They constitute a scheme as general as proportionality and correlation, which emerge simultaneously after the age of 12 to 13 (the formal operation stage in Piagetian theory). Combinatorial capacity is fundamental for hypothetical-deductive reasoning, which operates by combining and evaluating the possibilities in each situation. According to these authors, adolescents spontaneously discover systematic procedures of combinatorial enumeration, although for the permutations, it is necessary to wait until they are 15 years old.

More recent results, such as Fischbein's (1975), show that combinatorial problem solving capacity is not always reached, not even at that late age, without specific teaching. However, Fischbein and Gazit (1988) studied the effect of specific instruction on combinatorial capacity, and discovered that even 10-year-old pupils can learn some combinatorial ideas with the help of the tree diagram. Engel et al. (1976) have also been successful in teaching combinatorics to very young children and describe different games and activities to introduce this topic.

The role of problem solving in assessment

According to recent trends in mathematics education, mathematics is not just a symbolic language and a conceptual system, but mainly a human activity involving the solution of socially shared problems. The vision of the Curriculum and Evaluation Standards is that mathematical reasoning, problem solving, communication, and connections must be central in teaching and assessment. As stated in Garfield (1994), it is no longer appropriate to assess students' knowledge by having students compute answers and apply formulas. According to Romberg (1993), an authentic assessment should be developed by determining the extension in which the student has increased his/her ability to solve non-routine problems, reason, communicate, and apply mathematical ideas in a variety of related problems.

Consequently, both the teaching and assessment of combinatorics should be based on solving different combinatorial problems in which students need systematic enumeration procedures, recurrence, classification, tables, and tree diagrams. For example, let us consider the following item taken from Green's research concerning 11 to 16 year-old students' probabilistic reasoning (Green, 1981):

Example 2:

> Three boys are sent to the headmaster for cheating. They have to line up in a row outside the headmaster's room and wait for their punishment. No one wants to be first, of course! Suppose the boys are called Andrew, Burt, and Charles (A, B, C for short). We want you to write down all the possible orders in which they could line up. How many ways can the boys be lined up in?

Some students may use a tree diagram to write down all the solutions. In other cases, they may locate the boy to be lined up in first place (for example, Andrew), so that they reduce the problem to listing all the permutations of the two remaining boys, using recurrence (solving the problem with the help of a simpler version of the original problem). Even if some students proceed by trial and error, by writing down the possible permutations of the three boys without any systematic procedure, they may try to classify all the permutations produced according to the boy lined up in the first place, to check whether there is any forgotten permutation.

In this item, Green increased to four and five the number of boys to be lined up, to check whether students used recurrence and generalized their first solution to new parameter values. Using this item with our students, we found a typical mistake, consisting of giving solutions 8 and 10 for the permutations of four and five elements, after finding the correct number of permutations for using three-elements enumeration. Such reasoning is similar to that of the following student:

> Teacher: You have written ABC/ BCA/ CAB/ ACB/ BAC/ CBA for the different ways the three boys may be lined up. Why do you think there are just 8 different ways for the four boys to be lined up?

Student 1: As I have one more boy, I can put him in first or last place. So I have six and two that is equal to eight different ways.
Teacher: I see..., but, What about the five boys?
Student 1: Because you have eight ways with four boys, when you add another new boy, you can place him in first or last place, so you must add two to the eight possible ways and you obtain ten different ways.

To solve counting problems, students could start by enumerating some cases to discover the problem structure with or without the help of a tree diagram. They might use multiplication to count grouped collections instead, by formulating and applying addition and multiplication rules. As an example, we shall analyze the solution given by a 14-year-old girl, who had not been taught combinations and arrangement formulae, to the following counting problem:

Example 3:

Four children: Alice, Bert, Carol, and Diana go to spend the night at their grandmother's home. She has two different rooms available (one on the ground floor and another upstairs) in which she could put some or all the children to sleep. In how many different ways can the grandmother put the children into the two different rooms? (She could put all the children in one room). For example, she could put Alice, Bert and Carol into the ground floor room and Diana in the upstairs room.

This girl (Jessica) enumerated all the ordered decomposition of the number 4 into two addends (a partition of the four children in two groups), that is, $4=4+0 = 3+1= 2+2=1+3=0+4$. Then she counted the number of possibilities for each of these decompositions, and, finally, she applied the addition rule. If the students had been taught the combinatorial operation, they might have recognized the combinatorial operation that is the solution to the problem, here $AR_{2,4}$, i.e., the arrangements with replacement of two elements (the available rooms) taken four at a time (one room for each child).

Besides solving verbal problems, it is possible to ask students to prove combinatorial statements or to generalize the solutions found for a given problem. For example, we could ask the students to generalize the solution to Example 2, when there were n boys to be lined up. Both instruction and assessment should emphasize combinatorial reasoning as opposed to the application of analytic formulae for permutations and combinations. Instead of establishing the identity $C_{n,r}= C_{n,n-r}$ by algebraic manipulations, it is preferable for the students to reason that if you take a sample of r objects from n given objects there are still $n-r$ objects left. If we encourage students to formulate their own problems, we also will improve the quality of instruction as recommended in the NCTM (1991). On the other hand, Gal and Ginsburg (1994) noticed that the creation of a problem-solving environment requires an emotionally supportive atmosphere, where students feel safe to explore, are motivated to work longer, feel comfortable with temporary mistakes and are not afraid to apply different tools to the same problem, so we should try not to be too rigid in our teaching methods.

CLASSIFICATION OF COMBINATORIAL PROBLEMS AND IMPLICATIONS FOR TEACHING AND ASSESSMENT

As Webb (1993) stated, the interpretation of a student's responses implies making inferences about what a student knows. The items in the assessment instruments form a sample of the possible tasks concerning a specific concept or procedure, so obtaining more accurate inferences requires drawing as much varied information as possible. In this section, we analyze the main task variables in combinatorial problems in order to help teachers when selecting representative samples of problems for teaching and assessment purposes.

Implicit mathematical model in simple combinatorial problems

According to Dubois (1984), simple combinatorial configurations may be classified into three models: *selections*, which emphasize the concept of sampling; *distributions*, related to the concept of mapping; and *partition* or division of a set into subsets.

The selection model

In the *selection model*, a set of m (usually distinct) objects are considered, from which a sample of n elements must be taken, as asked in the following problem (Fischbein & Gazit, 1988):

Example 4:

> There are four numbered marbles in a box (with the digits 2, 4, 7, 9). We choose a marble and note its number. Then we put the marble back into the box. We repeat the process until we form a three-digit number. How many different three-digit numbers is it possible to obtain? For example, the number 222 is a possible combination.

The keyword "choose," included in the statement of the problem, suggests to the student the idea of sampling marbles from a box. Other key verbs that usually refer to the idea of sampling are "select," "take," "draw," "gather," "pick," etc.

For the student of probability it is easy to model counting methods by performing n drawings of m numbered balls from an urn. In selecting a sample, sometimes students are allowed to repeat one or more elements in the sample, as in example 4, and other times they are not. According to this feature and whether the order in which the sample is drawn is relevant (example 4) or not, we obtain four basic sampling procedures: a) with replacement and with order ($AR_{m,n}$), b) with replacement and without order ($CR_{m,n}$), c) without replacement and with order ($A_{m,n}$) and d) without replacement and without order ($C_{m,n}$) (permutations are a particular case of arrangements when $m=n$).

The distribution model

A second type of problem refers to the *distribution* of a set of n objects into m cells, such as in the following problem, in which each of the three identical cards must be introduced (placed) into one of four different envelopes (Batanero et al., to be published):

Example 5:

Supposing we have three identical letters, and we want to place them into four different colored envelopes: yellow, blue, red, and green. It is only possible to introduce one letter into each different envelope. How many ways can the three identical letters be placed into the four different envelopes? For example, we could introduce a letter into the yellow envelope, another into the blue envelope, and the last letter into the green envelope.

Other key verbs that could be interpreted in the distribution model are "place," "introduce," "assign," "store," etc. The solution to this problem is $C_{4,3}$, but there are many different possibilities in this model, depending on the following features:

- Whether the objects to be distributed are identical (as in this problem) or not.
- Whether the containers are identical or not, as in the example.
- Whether we must order the objects placed into the containers (this makes no sense in example 5 since the objects are identical).
- The conditions that you add to the distribution, such as the maximum number of objects in each cell, or the possibility of having empty cells and so on. (In the problem proposed you may only introduce one letter into each envelope and there is an envelope left empty, but these conditions could be changed.)

Assigning the n objects to the m cells is equivalent, from a mathematical point of view, to establishing an application from the set of the n objects to the set of the m cells. For injective applications we obtain the arrangements; in the case of a bijection we obtain the permutations. Nevertheless, there is no direct definition for the combinations using the idea of application. Moreover, if we consider a non-injective application, we could obtain a problem for which the solution is not a basic combinatorial operation, so there is not a different combinatorial operation for each different possible distribution. For example, if we consider the non-ordered distribution of n different objects into m identical cells, we obtain the second kind Stirling numbers $S_{n,m}$. Consequently, it is not possible to translate each distribution problem into a sampling problem. The reader may find a comprehensive study of Stirling numbers in Grimaldi (1989) and for the different possibilities in the distribution model in Dubois (1984).

The partition model

Finally, we might be interested in splitting a set of n objects into m subsets, i.e., performing a *partition* of the set, as in the following problem (Batanero et al., in press):

Example 6:

Mary and Cindy have four stamps numbered from 1 to 4. They decide to share the stamps, two for each of them. In how many ways can they share the stamps? For example, Mary could keep the stamps numbered 1 and 2 and Cindy the stamps numbered 3 and 4.

We may visualize the distribution of n objects into m cells as the partition of a set of m elements into n subsets (the cells).Therefore, there is a bijective correspondence between the models of partition and distribution, though for the student this may not be evident. Other key verbs associated with partition are "divide," "distribute," "split," "decompose," "separate," etc.

In our research, (Batanero et al., to be published) we showed that the three types of problems we have described (selections, distributions, and partition) are not equivalent in difficulty for the students, even after being taught combinatorics. Other task variables that have affected students' responses in Fischbein and Gazit's research (1988) are the combinatorial operation involved in the problem (combination, permutation, or arrangement), the sizes of parameters m and n, and the type of element to be combined (letter, numbers, people, objects).

Consequently, all these problem features should be considered as fundamental task variables in teaching and assessing combinatorics. As regards the sampling model, Hawkins et al. (1992) suggested that the attempt to describe a particular combinatorial problem by one of the sampling models will force the student to look carefully at the mechanism underlying random experiments. This proposal ought to be extended to the distribution and partition models. Suggesting that students conceptualize probabilistic and combinatorial problems using these three prototype models (selection, distribution, and partition) may not guarantee that the counting will be correct, but it will prevent some probabilistic misconceptions amongst the students. It may also help students to develop probabilistic reasoning, problem solving, heuristic strategies, communication and connections with other mathematical ideas, suggested as central points in the mathematics curriculum.

ASSESSING STUDENTS' DIFFICULTIES IN SOLVING COMBINATORIAL PROBLEMS

New assessment approaches are intended to better capture how students think, reason, and apply their learning. This requires focusing the problem of assessing mathematical knowledge from a new perspective, as "the comprehensive accounting of an individual's or group's functioning within mathematics or in the application of mathematics" (Webb, 1992, p. 662). The goal is assessing the implied processes and not only measuring the degree to which students have acquired a given content. A wider range of measures, most of them qualitative ones, would be needed (Romberg et al., 1991). Assessment is not the aim of educational experiments but rather a continuous and dynamic process that can be used by teachers to help students attain curricular goals. Therefore, a key point in assessing combinatorial reasoning is identifying the students' difficulties in solving combinatorial problems, some of which shall be described in this section.

Non-systematic enumeration

This difficulty consists of trying to solve the problem by enumeration using trial and error, without any recursive procedure leading to the formation of all the possibilities. Consider, for example, the interview with student 2 (15-years-old, who had not studied combinatorics yet) to explain his solution to the following problem (Batanero et al., 1994):

Example 7:

The garage in Angel's building has five spaces numbered from 1 to 5. Because the building is very new, there are only three different residents: Angel, Beatrice, and Carmen (A, B, and C) who park their cars in the garage. For example, Angel could park his car in place number 1, Beatrice in place number 2, and Carmen in place number 4. In how many different ways could Angel, Beatrice, and Carmen park their cars in the garage?

Teacher: How would you solve this problem?
Student 2: We have three cars: A, B, C, don't we? So I can park Angel's car in the first space, Beatrice's car in the second, and Carmen's in the third, so I write down A=1, B=2, C=3. Then, another position could be that I put Angel's car in the second space: A=2, B=3, C=1, or, perhaps,
A=3, B=1, C=2;
A=1, B=3, C=2;
A=2, B=1, C=3;
A=4, B=3, C=1;
A=3, B=1, C=4;
A=1, C=4, B=3;
A=3, C=4, B=1.
Teacher: Do you think there is any other possibility?
Student 2: I don't know... I suppose I could continue in different positions, because I haven't used 1, 2 and 4 yet, and I could change the order. What I mean is that it is not the same thing when Angel put his car in place number 2 as when it is Carmen who parks her car in space number 2. Do you see what I mean?

In spite of understanding the type of combinatorial configuration he was asked to produce, this student was unable to find all the different possibilities, because he did not follow a systematic procedure. To assess if he could solve the problem with a smaller value of the parameters, we asked him to solve the same problem with only two people (Angel and Beatrice) and three spaces in the garage. Below we reproduce what the student wrote as the solution to this new problem:

A=1, B=2;
A=2, B=1;
A=1, B=3;
A=3, B=1;
A=2, B=3;
A=3, B=2.

With this smaller number of elements the student followed a system. Nevertheless, he was unable to use recurrence to link the original problem of parking three cars with the solution obtained for this simpler version.

Incorrect use of the tree diagram

Tree graphs are one of the most useful resources for visualizing both combinatoric and probabilistic situations. In Fischbein's terminology they belong to "diagramatic models" and present important intuitive characteristics. They offer a global representation of the situation structure and this contributes to the immediacy of understanding and to finding the problem solution. In spite of this importance, Pesci (1994) proved that students found it difficult to build suitable tree diagrams to represent problem situations and, so, the same graph is the cause of many errors.

Error of order

This mistake consists of confusing the criteria for combinations and arrangements, i.e., distinguishing the order of the elements when it is irrelevant or on the contrary, not considering the order when it is essential. Here is one example taken from a student's written solution to the following example.

Example 8:

In how many ways can a teacher select three pupils to rub out the blackboard, if five students (Elisabeth, Ferdinand, George, Lucy, and Mary) have offered to do it?

E= Elisabeth, F= Ferdinand, G= George, L= Lucy, and M=Mary
E F G, E F L, E F M, E G F, E G L, E G M, E LM, E L F, E L G, E M F, E M G, E M L;
12 x5= 60; therefore, you have 60 different ways.

Error of repetition

The student does not consider the possibility of repeating the elements when it is possible, or he/she repeats the elements when there is no possibility of doing so. This is an example in a student's written answer to the following item, adapted from Fischbein and Gazit (1988), in which we also note the lack of systematic enumeration:

Example 9 :

In an urn there are three marbles numbered with the digits 2, 4, and 7. We extract a marble from the urn and note its color. Without replacing the first marble, we extract another one and note its number. Finally, we extract the last marble from the urn. How many three-digit numbers can we obtain with this method? For example, we could obtain the number 724.

724, 742, 722, 772, 744, 472, 427, 477, 444, 422, 274, 247, 277, 222, 244;
you can have 15 different numbers.

Confusing the type of object

This type of error occurs when students consider that identical objects are distinguishable or that different objects are indistinguishable. Below we reproduce the interview that we carried out with a student about his solution to the following problem (Fischbein et al., 1970):

Example 10:

Each one of five cards has a letter: A, B, C, C and C. In how many different ways can I form a row by placing the five cards on the table? For example, I could place the cards in the following way: ACBCC.

Student 3: It is a permutation of five elements without repetition.
Teacher: Why do you think it is a permutation?
Student 3: I do not remember very well... It is the same thing as when you need to place five books on a shelf, because you cannot repeat the book. The different ways in which you may line up the letters. That would be permutation.

The teacher asked the student to write down all the different possibilities for placing the three letters ACC on the table, thereby reducing the size of the parameters in order to simplify the problem and to better understand the student's reasoning. The student started writing:

Student 3: AC_1C_2.
Teacher: Why do you write C_1C_2? I wanted you to use A, C and C!
Student 3: But I need to differentiate between the two Cs, because you have two different cards, although each has the same letter C. So, these are all the possibilities: AC_1C_2, AC_2C_1, C_1AC_2, C_2AC_1, C_1C_2A, C_2C_1A; there are six in total.

Confusing the type of cell (the type of subsets) in partition or distribution models

This mistake consists of believing that we could distinguish identical (subsets) cells, or that it is not possible to differentiate the distinguishable cells (subsets). For example, in the following item some students only consider the three different ways in which the set of the four students can be divided into two groups. So they do not differentiate which group was going to complete the mathematics project and which was going to undertake the language project.

Example 11:

Four friends: Ann, Beatrice, Cathy, and David, must complete two different projects: one in mathematics, and the other one in language. They decided to split up into two groups of two pupils, so that each group could perform a project. In how many different ways can the group of four pupils be separated to perform these projects? For example, Ann and Cathy could complete the mathematics project, and Beatrice and David the language project.

Misunderstanding the type of partition required

This can occur in the following two ways: The union of all the subsets in a partition does not contain all the elements of the total set, or some possible partitions are forgotten. We can observe these two errors in the answer provided by a student to the following problem, in which he only considered two types of partitions: giving all the cars to one child or giving only one car to each child.

Example 12:

A girl has four different colored cars (black, orange, white, and gray) and she decides to share the cars with her brother John, and her sisters Peggy and Linda. In how many different ways can she share the cars? For example, she could give all the cars to Linda.

Student 4: black, orange, white, gray for Peggy;
black, orange, white, gray for John;
black, orange, white, gray for Linda;
black for Peggy; black for John; black for Linda;
orange for Peggy; orange for John; orange for Linda;
white for Peggy; white for John; white for Linda;
gray for Peggy; gray for John; gray for Linda.

IMPLICATIONS

In this chapter we have shown that combinatorial reasoning is not restricted to solving verbal combination and arrangement problems, but that it includes a wide range of concepts and problem-solving abilities. Most of these components are fundamental tools in developing probabilistic reasoning and in attaining the curricular probabilistic goals for primary and secondary education. With the help of manipulative materials and tree diagrams, meaningful activities linked to probability may be proposed, even to very young children. These activities may also serve to develop and assess problem solving and communication skills and connections to other mathematical topics.

Some of the task variables we have described in this chapter, especially the implicit mathematical model, have shown their strong effect on both the difficulties of the combinatorial problems and the types of errors in different research work. Consequently, we need to consider these task variables when we assess students' combinatorial reasoning if we want to get a more comprehensive idea of students' capabilities and conceptions.

These variables also need to be recognized when organizing our teaching, which should also emphasize the modeling process, the recursive reasoning and the systematic procedures of enumeration, instead of merely concentrating on algorithmic aspects and on the definitions of the combinatorial operations.

We have also presented some examples of tasks used to assess combinatorial reasoning in different experimental research, which we proposed to students in written questionnaires or in interviews with and without the help of manipulative material. Other teachers may want to use our examples to build their own items for teaching or assessment purposes, changing the values for task variables as needed. Or, our various examples could be included in different assessment methods, such as questions, exams, homework, portfolios, interviews, classroom discussion, and individual or collective projects (Garfield, 1994). As stated by Webb (1993), any form of assessment includes not just the task, but the students' responses, the interpretation of these responses, the meaning given to them, and the report on the assessment findings.

The careful reporting and analysis of our students' responses is an essential part of our success as teachers. Of particular benefit to our understanding of student difficulties with combinatorial reasoning is the classification of such responses into clearly defined categories. A thorough appreciation of the information in this chapter will assist us as teachers as we assign suitable meanings to our students' progress in combinatorics and probability.

Chapter 19

Probability Distributions, Assessment and Instructional Software: Lessons Learned from an Evaluation of Curricular Software

Steve Cohen and Richard A. Chechile

Purpose

While mathematics education guidelines have encouraged substantial change in the introductory probability and statistics curriculum, probability distributions still remain an important topic in a first course. In fact, just as software has made data analysis more accessible to students in introductory courses, it also offers new ways to teach probability distributions. However, these new teaching technologies, which emphasize active experimentation and interpretation of displays, also raise new questions. Just what do students see when they examine a display of a probability distribution? Do the displays really help students acquire a clear conceptual understanding? Can interactive exercises for related concepts like sampling distributions make good use of displays? Finally, can good assessment practices help us learn when displays are effective and when they might be confusing? This chapter will discuss some interactive, computer-based exercises that use and teach probability distributions, and consider how assessment can help address some of the important questions these new teaching technologies raise.

INTRODUCTION

The past decade has produced a great deal of thinking about the curricular goals for introductory courses in probability and statistics. Some of the thinking and research has been designed to demonstrate how deep the educational problems run. Garfield and Ahlgren (1988) have pointed out just how hard certain concepts are for students. Shaughnessy (1992), in a review of research on probability and statistics education, confirmed their findings. As a result of this research, calls for new instructional goals and methods of teaching have begun to redefine the introductory statistics curriculum. Cobb (1992) outlined several recommended changes. The recommendations included an emphasis on (a) statistical thinking rather than formulas and cookbook approaches, (b) more hands-on work with data, and (c) the use of interactive technologies to make data analysis more efficient and to help teach difficult concepts.

However, one traditional part of some introductory courses in probability and statistics has been a bit lost in this discussion. The role of the probability distribution in learning and teaching has been overlooked. In spite of the emphasis placed on hands-on data analysis and alternative methods for inference, probability distributions are, and likely will always be, a major part of a first course. Why? To begin, probability distributions, like the normal curve, are not found only in statistics courses and text books. They are part of the vocabulary for communicating basic ideas. Normal probability distributions are used to describe basic phenomena like grades and intelligence. While they might not be quite as ubiquitous as more common statistical concepts like averages and frequencies, they clearly extend beyond the classroom and consequently deserve a special place in the statistics curriculum.

Unlike data, probability distributions are formal theoretical models that describe the likelihood of a variable taking on a value (i.e., the binomial distribution) or a range of values (i.e., the normal distribution). They are useful for describing the behavior of sample means (i.e., a theoretical sampling distribution) and as such are important for teaching concepts (Moore, 1990) such as the central limit theorem and confidence intervals, as well as for doing basic statistics (NCTM, 1989). Their theoretical nature makes probability distributions a natural contrast to data which may help students develop a notion of stochasm. Understanding the *differences* between a data distribution and a probability distribution is one of the most profound insights a student can have. Indeed, the NCTM Curriculum and Evaluation Standards for School Mathematics (1989) encourages students in grades 5-8 to compare data with formal models to get a sense of stochasm.

Thus, even with a new, well placed emphasis on working with data, probability distributions still play an important part in teaching and learning basic statistics. Being able to interpret a probability distribution, and make well-reasoned claims about a variable by studying its probability distribution, is an important skill for almost any student in a first course in probability and statistics. How can we learn if we are successfully helping students achieve this goal?

INTERACTIVE EXERCISES INVOLVING PROBABILITY DISTRIBUTIONS

While all probability distributions have a mathematical form, only the most sophisticated students typically see, or choose to see, the associated equation. A more realistic goal, and one consistent with the kinds of changes called for by Cobb (1992), is to have students be able to interpret a display of a probability distribution and understand how it conveys probability.

One medium particularly well suited to helping students achieve these goals is statistical or curricular software. Software offers an excellent means to display probability distributions and to have students experiment with their different properties. Exercises may be focused on familiarizing students with properties of distributions themselves, or might use probability distributions to help illustrate other related concepts such as the central limit theorem. For instance, Figure 1 shows an interactive exercise from a program called ConStatS. The exercise is designed to help students understand the relationship between the parameters of a normal distribution and the shapes the distribution can take on. Students can change the parameters, and the resulting distribution is plotted on the same set of axes as the original distribution.

When displayed, probability distributions look deceptively simple. A set of probabilistic concepts is truly hidden in the plot of a simple function. It is easy to assume that students grasp the implied meaning. As a goal of an introductory course, it is important to know if a student can work with a probability distribution and see in them the rich and difficult concept of probability. This assessment goal is especially important if distributions are included as educational aids in other exercises (i.e., teaching sampling distributions).

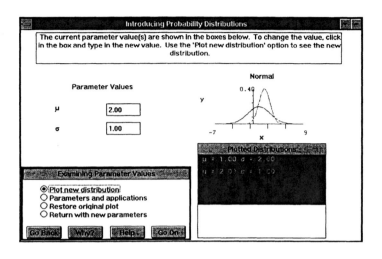

Figure 1. Exercise from ConStatS

ASSESSING STUDENTS' UNDERSTANDING OF PROBABILITY DISTRIBUTIONS

The remainder of this chapter will look at examples of instructional technology that teach or use probability distributions and consider the implications for assessing student understanding. In some instances the focus will be on learning probability distributions, and in other instances probability distributions will be used in examples designed to illustrate challenges in statistics assessment. Examples will be drawn from ConStatS — a set of interactive programs for helping students conceptualize topics covered in an introductory course (Cohen, Smith, Chechile, & Cook, 1994). Many of the insights in this chapter come from a three year, multi-site study of ConStatS (Cohen, Chechile, Smith, Tsai, & Burns, 1994). The scope of the evaluation (16 different introductory statistics courses at five different universities) should permit the results to generalize to most introductory courses and to most software for illustrating statistical concepts.

The chapter will proceed by first presenting some interactive educational exercises from ConStatS involving probability distributions. We will first look at the exercises and consider the educational goal(s). Then we will turn to the design and interpretation of assessment items, and look at how student performance on those items relates to overall conceptual understanding of material taught in a first course. The chapter will conclude with remarks about how assessing students' understanding of probability distributions connects to the bigger picture of assessment.

What is the software trying to teach?

ConStatS was created to help students gain a deep conceptual understanding of topics taught in introductory probability and statistics. This goal narrowed the assessment objectives in several important ways (Cohen et al., 1994). There is no emphasis in ConStatS on computational proficiency or remedial mathematical issues. For example, while the exercise shown in Figure 1 encouraged students to experiment with the parameters of various distributions and investigate the resulting forms, they were never required to calculate specific probabilities or become familiar with the mathematical form of the equation (though this is an option in the program). Consequently, almost all assessment items were designed exclusively to test conceptual understanding.

Yet even this focused educational goal permits a variety of questions that assess what students might have learned from experimenting with parameters of a distribution. For instance, Figure 2 shows a question used to assess a student's conceptual understanding of how the assigned mean and variance of a normal distribution affect its form. While this particular question tests conceptual understanding, other types of questions could assess similar kinds understanding. Specifically the question in Figure 2 is a *production* question. It requires a student to sketch a normal distribution with a given set of parameters. Alternatively, the question could have been designed to assess only *comprehension* by showing two distributions and asking the student to estimate the parameter values or to select from a set of possible options. It is important to look closely at the instructional exercise and consider what students are empowered to do by the software. The specific kinds of interactions can limit (or extend) learning, and the assessment items needs to reflect this. We will return to this issue later in the chapter.

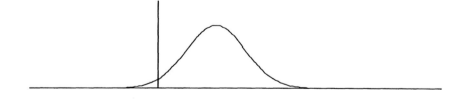

Below is a normal probability distribution with mean = 2 and variance =1.
On the graph, sketch a normal distribution with mean = 1 and variance = 2.

Figure 2. Item assessing understanding of the mean

USING PROBABILITY DISTRIBUTIONS TO HELP TEACH OTHER CONCEPTS

Some ConStatS programs use probability distributions to help teach other concepts. The two separate kinds of environments within ConStatS–programs treating probability distributions directly and those using them in context–lead to very different kinds of learning and require separate kinds of assessment items. For instance, there are several sampling programs that require students to use and interpret probability distributions as both population distributions and sampling distributions. Figure 3 shows a portion of the sampling distributions program with a normal probability distribution representing the population of human baby weights. In this environment, students are required to interpret a probability distribution, not construct one. Students who can effectively interpret a probability distribution in the context of a sampling exercise may not have a useful grasp of probability distributions in other contexts. It is possible, for instance, that a student could use information about a random variable (in this case human baby weights) to interpret the probability distribution in the sampling distributions setting. They may know that newborns are typically around seven pounds, and rarely weigh more than ten pounds and less than four pounds. This knowledge is consistent with a correct interpretation of the normal population distribution shown. However, the same student could have trouble interpreting a distribution without supporting information (i.e., for an abstract variable like X). This assessment issue emerges when considering the goal of the sampling distributions exercise. Students may be able to use intuitions about baby weights to make some sense of the population distribution, but then find themselves lost when looking at the sampling distribution of means about which they have no intuitions.

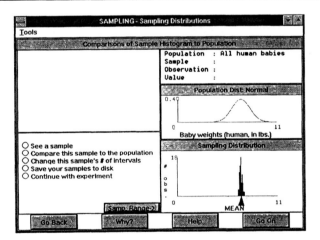

Figure 3. Sampling Distribution program

Exercises like the one in Figure 3 require students to interpret displays of probability distributions in order to learn statistical concepts like sampling distributions. Students who do not demonstrate a good understanding of spread when visually interpreting a normal distribution (i.e., do poorly on the question in Figure 2) may only partially grasp the concept of a sampling distribution. They may learn that sampling distributions describe distributions of statistics rather than individual data points, but may not come to understand the reduction in variance obtained when working with a sampling distribution rather than a population distribution.

The question in Figure 2 was included in the ConStatS evaluation. Students were typically able to shift the distribution correctly one unit to the left, but they had a great deal of trouble representing the adjusted variance (Cohen, Smith, Chechile, Burns & Tsai, 1996). The result illustrates the complex nature of using displays to convey concepts. Many students may have only a partial understanding of probability distributions. Ultimately, measuring a student's understanding of probability distributions is likely to require several well-formed questions that target specific properties and uses. Without such a detailed assessment, it may be impossible to learn why students have trouble with other key concepts like sampling distributions.

The effect of an interactive, visual medium

ConStatS, like most statistical software, makes extensive use of graphical representations. Many of the exercises in the software, like the one shown in Figure 1, require students to manipulate displays in order to learn and to interpret distributions. Even if students successfully learn how the parameters μ and σ influence normal probability distributions by doing experiments and studying resulting displays, the display may cause other confusions. For instance the normal distribution, as a theoretical construct, has a range from negative to positive infinity. The display, however, is limited. The range of the distribution may only be plotted to three or four standard deviations, and a student will only see the distribution(s) displayed over a finite range. The display tacitly conveys that the normal distribution is just a *finite*, smooth curve that represents some symmetrical data, rather than a distinct theoretical function that represents probability from negative to positive infinity. The limited display may reinforce this mistaken view. Figure 4 presents a way to assess student understanding about the range of the random variable. To supply a correct answer students must use a property of the normal probability distribution that is not evident from the display.

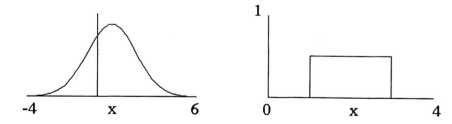

Figure 4. Item assessing the range of random variables

This assessment issue points to the potential limitations of using a visual and graphical medium to illustrate subtle ideas. Having students learn by investigating a representation of probability creates a special demand on assessment. It requires an assessment of the potential conceptual confusions that might emerge owing to the limits of a particular representation.

MATCHING ASSESSMENT ITEMS TO INSTRUCTIONAL MATERIALS

If assessments of new instructional methods like ConStatS are going to provide feedback on how the methods should be improved, the assessment item needs to match instructional method very closely. Earlier in the chapter the discussion centered on a question for assessing a student's understanding of the parameters for the normal distribution. Two different kinds of questions were considered: a production question and a comprehension question. It is difficult to determine which of these two kinds of questions is the best match for the exercise in the software. Students using the exercise in the software (Figure 1) first selected an existing distribution and then saw the values of the parameters. They were then asked to modify the parameters, and review the resulting distribution. In this exercise, students are (in some sense) producing a normal distribution by defining a new set of parameters and having the software draw the new form of the distribution. Owing partially to a limitation in the instructional method, students are never actually required to sketch or otherwise produce a distribution by hand (as required by the question). Consequently, some students may not confront the kinds of subtle decisions that sketching a graph demands. Having students sketch the graph may force them to pay much closer attention to inflection points, the total area covered, the range of the plot, and other more detailed properties of the distribution and the display. Sketching is a very different educational exercise than creating displays by computer, yielding a different set of skills and appreciation for the distribution. It is the kind of educational exercise that is often omitted when students move from pencil and paper exercises to software.

How does reconsideration of the exercise in Figure 1 influence interpretations of student responses to the question in Figure 2? In this instance, a production question might not match the instructional method quite as well as a comprehension question. Students who do not do well on the question in Figure 2 may have a gained a basic understanding of parameters and forms of the normal distribution from the exercise, enough to answer a comprehension question correctly. Yet their understanding may not generalize well to slightly more demanding tasks.

INTERPRETING OPEN-ENDED QUESTIONS

Production questions, like the one in Figure 2, raise the additional issue of the value of open-ended questions. When using a new instructional medium like software to help teach sophisticated concepts, open-ended questions can be very valuable. Students come to a question with a wide range of skills and knowledge and offer responses that are difficult to anticipate. Often these unanticipated responses offer clues to conceptual confusions. An example from the ConStatS evaluation should illustrate the point. The question in Figure 5 required students to construct a uniform probability density function that matched the cumulative density function. Several students drew "mostly" uniform distributions with curved tail areas from a normal distribution at the ends. It appears that at least some students have trouble abandoning a normal model when considering other probability distributions. Students who demonstrate a good conceptual understanding of normal distributions may display subtle misunderstandings when asked to interpret other probability distributions. Questions with these kinds of unanticipated answers may be very difficult to score, but they provide evidence about what kinds of misunderstandings plague students. When many students exhibit the same kind of misconception, it may indicate a strong cognitive bias that is difficult to address by instruction (Smith, diSessa, & Roschelle, 1992; Cohen, et al., 1996). Production questions allow students to demonstrate common misconceptions through patterns of errors that are often difficult to anticipate.

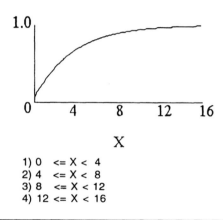

Figure 5. Probability density function

In many instances, it is possible to predict the kinds of mistakes students might make if they lack a certain understanding or have a certain misunderstanding. Under these circumstances, short answer questions that play on the misconceptions are very useful for separating students who have a good understanding from those who do not. Figure 6 shows a cumulative density function and requires students to indicate which range of values the variable is most likely to take on.

It is possible students might misinterpret the plot as a probability density function and select the option corresponding to the largest area under the curve. Another possibility is that students believe that the height of the ordinate, which represents probability, indicates the likelihood of any given value being observed (i.e., they read it as if it was a bivariate plot). Students with either of these misconceptions will almost certainly select the last option, 12<=x<=16. Unfortunately, the question does not help distinguish which of the anticipated misconceptions is serving as the chief impediment to successful interpretation of the graph. Items like the one in Figure 6 must be improved to provide more precise feedback about the particular misconceptions operating.

The questions in Figures 5 and 6 also illustrate another principle for evaluating conceptual understanding, transference. The key is to instantiate the concept under evaluation in a scenario unfamiliar to the student. Many students may be familiar with normal distributions, and with the relationship between the normal probability density function and its cumulative density function. They may have seen the relationship in a book, or even encountered it in another class. Their understanding may be limited to a rote visual recollection rather than a true understanding of the relationship. Using an unfamiliar distribution helps reduce the chance that a student is relying only on rote memory, rather than reasoning, to answer the question.

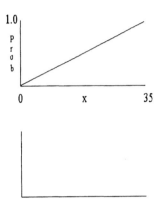

Construct a probability distribution which corresponds to the cumulative probability distribution shown below.

Figure 6. Cumulative density function

IMPLICATIONS

As the goals of teaching probability and statistics move from calculating numbers to reasoning and interpretation, and the teaching medium changes from blackboards to software, assessment practices must also adapt. The kind of assessment issues raised are not unique to ConStatS. Many computer programs used in the statistics curriculum, whether designed for doing statistics (like SPSS or Data Desk) or teaching statistics, use representations in similar ways. For instance, Rubin and Rosebery (1990) describe an interactive tool for the pre-college sector called ELASTIC, one part of which invites students to interactively stretch bars on histograms and examine changes in statistics. As in ConStatS, the pedagogy involves interacting with representations and drawing inferences about related statistical concepts. This pedagogy, and its educational goals, requires assessment practices that examine the components of the pedagogy and help us understand the effectiveness of representations and interactions. What questions should drive the construction of appropriate assessment strategies?

When considering exercises like the ones used in ConStatS, several issues emerge. When students are shown a probability distribution in an instructional exercise, what exactly do they see?

What concepts does it evoke? And what concepts must it evoke for students to use it in a learning exercise? Assessment items that provide reliable answers to these questions need to be crafted with a great deal of care. For instance, in the sampling distributions program pictured in Figure 3, the hope is that students, when seeing the population distribution, use concepts of probability to interpret sampling processes. The representation of a probability distribution needs to be a window into probabilistic concepts. While the set of concepts in a student's mind may not include all the formal properties of probability and probability distributions, they should permit the student to realize why a probabilistic definition of the variable is salient and how using the variable in a population setting is different from using it when it is described by sample data.

Do these visual, non-mathematical representations open the necessary conceptual window? At least two questions in the ConStatS assessment cast some doubt on whether displays of probability distributions always fulfill this role. In the question shown in Figure 4, students are asked to identify the highest and lowest values the variable described by the distribution could take on. Many students selected the range from the display, rather than true limits of the distribution. On another question, students were shown a normal population distribution that represents the weight of newborn cats and asked what the disadvantages of using a normal distribution might be to describe feline birth weight. Many students claimed that the normal distribution would *not account for outliers* (Cohen, et al., 1996). It appears that the limits of the display were taken to define the limits of observable data. This interpretation would be appropriate if students believed that they were looking at data. In this instance, the pattern of errors across questions indicates that some students may be confusing displays of data with displays of probability distributions. Why might this be happening?

There are several possible interpretations. Some students may not have well-formed ideas about probability, and interpret probability distributions as univariate data. It may also be that students do have separate, well-formed representations of probability and data, but probability distributions do not invoke the correct set of concepts. Finally, it may be that in spite of these errors, students do bring a suitable subset of probabilistic concepts to the instructional exercise. If either of the first two options is true, especially the second, then using probability distributions (or any abstract graphical representation) in an instructional exercise may be a delicate matter.

A principal goal of any assessment must be to help identify this kind of subtle confusion. If these subtle confusions can be identified and corrected, displays of probability distributions can be used to successfully illustrate a number of important concepts. The cumulative wisdom gained from good assessment will help teachers know when they can draw a probability distribution and have their students understand its meaning.

One final point to emphasize here is that results from two different kinds of questions supported each other in illustrating a misunderstanding. The question on high and low values requires only that students interpret a plot, while the question on population distributions for newborn kittens has a generative component (i.e., students must produce an explanation.) Assessment items and practices can and should complement each other. Different kinds of assessment items will help provide a more complete profile of what students are and are not learning, and why. There is no one correct venue for assessment. Student explanations, like those required for the question on population distributions, can come in informal settings, like computer labs, or as part of formal assessment items. Even short-answer multiple-choice questions, like the one in Figure 6, can probe conceptual understanding. What is important is to create a variety of measures that complement each other, and which offer opportunities to examine student understanding and improve curricula.

References

Allwood, C. M., & Montgomery, H. (1981). Knowledge and technique in statistical problem solving. *European Journal of Science Education, 3*, 431-450.
Allwood, C. M., & Montgomery, H. (1982). Detection of errors in statistical problem solving. *Scandinavian Journal of Psychology, 23*, 131-140.
American Association for the Advancement of Science (Project 2061) (1993). *Benchmarks for science literacy.* New York, NY: Oxford University Press.
American Statistical Association-National Council of Teachers of Mathematics (ASA-NCTM) (1994). *Teaching statistics: Guidelines for elementary through high-school.* (G. Burrill, Editor). Palo Alto, CA: Dale Seymour Publications.
Anderson, C. W. & Loynes, R. M. (1987). *The teaching of practical statistics.* Chichester & New York: John Wiley.
Anderson, S, Ball, S., Murphy, R., & Associates. (1975). *Encyclopedia of educational evaluation.* San Francisco: Jossey-Bass.
Arter, J. & Spandel, V. (1992). Using portfolios of student work in instruction and assessment. *Educational Measurement: Issues and Practice, 11*(1), 36-44.
Artzt, A. F. & Armour-Thomas, E. (1992). Development of a cognitive-metacognitive framework for protocol analysis of mathematical problem solving in small groups. *Cognition and Instruction, 9*, 137-175.
Artzt, A. F. (1994, February). Integrating writing and cooperative learning in the mathematics class. *Mathematics Teacher, 87*, 80-85.
Australian Education Council. (1991). *A national statement on mathematics for Australian schools.* Canberra: Curriculum Corporation.
Australian Education Council (1994a). *A statement on studies of society and environment for Australian schools.* Carlton, Vic.: Curriculum Corporation.
Australian Education Council (1994b). *Studies of society and environment – a curriculum profile for Australian schools.* Carlton, Vic.: Curriculum Corporation.
Baker, E. L. (1990). Developing comprehensive assessments of higher order thinking. In G. Kulm (Ed.), *Assessing higher order thinking in mathematics* (pp. 7-20). Washington, DC: American Association for the Advancement of Science.
Baker, E. L., Aschbacher, P. R., Niemi, D., & Sato, E. (1992). CRESST Performance assessment models: Assessing content area explanations. Los Angeles: CRESST.
Baker, E. L., Freeman, M., & Clayton, S. (1991). Cognitively sensitive assessment of subject matter. In M. C. Wittrock & E. L. Baker (Eds.), *Testing and cognition.* New York: Prentice-Hall.
Batanero, C., Godino, J. D. & Navarro-Pelayo, V. (1994). *Razonamiento combinatorio* [Combinatorial reasoning]. Madrid: Síntesis.

Batanero, C., Navarro-Pelayo, V. & Godino, J. (In press). Effect of the implicit combinatorial model on combinatorial reasoning of secondary school students. *Educational Studies in Mathematics.*

Begg, A. (1991). Assessment and constructivism. *New Zealand Mathematics Magazine. 28*(2), 14-20.

Ben-Zvi, D., & Friedlander, A. (1996). Statistical thinking in a technological environment. In C. Batanero (Ed.), *Proceedings for the International Association for Statistical Education roundtable on research on the role of technology in teaching and learning statistics* (pp. 57-67). Spain: University of Granada.

Bentz, H. J., & Borovcnik, M. G. (1988). *Empirical research on probability concepts.* University of Klagenfurt, Austria.

Biggs, J. B., & Collis, K. F. (1982). *Evaluating the quality of learning: The SOLO taxonomy.* New York: Academic Press.

Biggs, J. B., & Collis, K. F. (1991). Multimodal learning and the quality of intelligent behaviour. In H. A. H. Rowe (Ed.), *Intelligence: Reconceptualisation and measurement* (pp. 57-76). Hillsdale, NJ: Lawrence Erlbaum.

Brandt, R. (1992). On performance assessment: A conversation with Grant Wiggins. *Educational Leadership, 49*(8), 35-37.

Bransford, J., Hasselbring, T., Barron, B., Kulewicz, S., Littlefield, J., & Goin, L. (1989). Uses of macro-contexts to facilitate mathematical thinking. In R. I. Charles & E. A. Silver (Eds.), *The teaching and assessing of mathematical problem solving* (pp. 125-147). Reston, VA: National Council of Teachers of Mathematics.

Bransford, J. & Vye, N. (1989). A perspective on cognitive research and its implications in instruction. In L. B. Resnick & L. E. Klopfer (Eds.), *Toward the thinking curriculum: Current cognitive research* (1989 Yearbook of the Association for Supervision and Curriculum Development). Alexandria, VA: ASCD.

Bright, G. W., & Friel, S. N. (in press). Graphical representations: Helping students interpret data. To appear in S. P. Lajoie (Ed.), *Reflections on statistics: Agendas for learning, teaching, and assessment in K-12.* Hillsdale, NJ: Lawrence Erlbaum.

Bright, G., & Hoeffner, K. (1993). Measurement, probability, statistics, and graphing. In D. T. Owens (Ed.), *Research Ideas for the Classroom: Middle Grades Mathematics* (pp. 78-98). New York: MacMillan Publishing Co.

Carnevale, A. P., Gainer, L. J. & Meltzer, A. S. (1990). *Workplace basics: The essential skills employers want.* San Francisco: Jossey-Bass.

Carpenter, T., & Lehrer, R. (in preparation). Learning mathematics with understanding. In E. Fennema & T. Romberg (Eds.), *Classrooms that Promote Understanding.*

Case, R. (1985). *Intellectual development: Birth to adulthood.* New York: Academic Press.

Castles, I. (1992). *Surviving statistics: A user's guide to the basics.* Canberra: Australian Bureau of Statistics.

Charles, R., Lester, F., & O'Daffer, P. (1987). *How to evaluate progress in problem solving.* Reston, VA: National Council of Teachers of Mathematics.

Clark, M. (1994). The effect of context on the teaching of statistics at first year university level. In L. Brunelli & G. Cicchitelli (Eds.), *IASE, Proceedings of the First Scientific Meeting* (pp. 105-113). Perugia, Italy: University of Perugia.

Clayden, A. D., & Croft, M. R. (1989). *Statistical consultation - Who's the expert?* Paper presented at the AI and Statistics Conference, Fort Lauderdale, Florida.

Cobb, G. W. (1992) Teaching statistics. In L. Steen (Ed.), *Heeding the call for change: Suggestions for curricular action.* Washington, DC: The Mathematical Association of America.

Cockcroft, W. (chair) (1982). *Mathematics counts.* London: Her Majesty's Stationery Office.

Cohen, S., Chechile, R., Smith, G., Tsai, F., & Burns, G. (1994). A method for evaluating the effectiveness of educational software. *Behavior Research Methods, Instruments & Computers. 26* (2), 236-241.

Cohen, S., Smith, G. E., Chechile, R. A., Burns, G., & Tsai, F. (1996) Impediments to learning probability and statistics identified from an evaluation of instructional software. *Journal of Educational and Behavioral Statistics. 21* (1), 35-54.

Cohen, S., Smith, G., Chechile, R., & Cook, R,. (1994) Designing software for conceptualizing statistics. In L. Brunelli & G. Cicchitelli (Eds.), *IASE: Proceedings of the First Scientific Meeting* (pp. 237-245). Perugia, Italy: University of Perugia.

Collins, A., Brown, J. S., & Newman, S. E. (1989). Cognitive apprenticeship: Teaching the craft of reading, writing, and mathematics. In L. Resnick (Ed.), *Knowing, learning, and instruction: Essays in honor of Robert Glaser* (pp. 453-494). Hillsdale, NJ: Lawrence Erlbaum.

COMAP, 1990. *Against all Odds: Inside statistics.* A publication of the Consortium for Mathematics and Its applications. New York: W. H. Freeman.

Cooper, W. & Davies, J. (Eds.) (1990 to present). *Portfolio Assessment Clearinghouse Newsletter.* (Available from San Dieguito Union High School district, 710 Encinitas Blvd., Encinitas, CA 92024, 619-753-6491).

Curcio, F. R. (1981). *The effect of prior knowledge, reading and mathematics achievement, and sex on comprehending mathematical relationships expressed in graphs.* (Final report to the National Institute of Education). Brooklyn, NY: St. Francis College. (ERIC Document Reproduction Service No. ED 210 185)

Curcio, F. R. (1987). Comprehension of mathematical relationships expressed in graphs. *Journal for Research in Mathematics Education, 18,* 382-393.

Curcio, F. R. (1989). *Developing graph comprehension: Elementary and middle school activities.* Reston, VA: National Council of Teachers of Mathematics.

Curcio, F. R. & Artzt, A. F. (in review). Students communicating in small groups: Making sense of data in graphical form. In M. Bartolini-Bussi, A. Sierpinska, & H. Steinbring (Eds.), *Language and communication in the mathematics classroom.*

Dallal, G. E. (1990). Statistical computing packages: Dare we abandon their teaching to others? *The American Statistician, 44*(4), 39-42.

Dauphinee, T. L., Schau, C., & Stevens, J. J. (in press). Survey of Attitudes Toward Statistics: Factor structure and factorial invariance for females and males. *Structural Equation Modeling.*

Davis, R. B., Maher, C. A., & Noddings, N. (1990). Constructivist views of the teaching and learning of mathematics. *Journal for Research in Mathematics Education*, Monograph 4. Reston, VA: National Council of Teachers of Mathematics.

Decriminalise drug use: poll.(1992, September 26). *Hobart Mercury,* p. 3.

Del Vecchio, A. M. (1994). *A psychological model of statistics course completion.* Unpublished doctoral dissertation, University of New Mexico, Albuquerque.

Dossey, J. A., Mullis, I. V. S., & Jones, C. O. (1993). *Can students do mathematical problem solving?* Washington, DC: US Department of Education.

Dubois, J. G. (1984). Une systematique des configurations combinatoires simples [A systematic for simple combinatorial configurations]. *Educational Studies in Mathematics, 15,* 37-57.

Edwards, K. (1990). The interplay of affect and cognition in attitude formation and change. *Journal of Personality and Social Psychology, 59*, 202-216.
Elliott, G. D. & Starkings, S. A. (1994). *Project assessment criteria*. London, England: South Bank University.
Engel, A., Varga, T. & Walser, W. (1976). *Hasard ou strategie?* [Chance or strategy?]. Paris: O.C.D.L.
English, F. (1992). *Deciding what to teach and test*. Newbury Park, CA: Corwin Press.
Ewing, T. (1994, January 6). Scientists urge guard against comet disaster. *The Melbourne Age*, 3.
Fischbein, E. (1975). *The intuitive sources of probabilistic thinking in children*. Dordrecht, The Netherlands: Reidel.
Fischbein, E. (1987). *Intuition in science and mathematics*. Dordrecht, The Netherlands: Reidel.
Fischbein, E. & Gazit, A. (1988). The combinatorial solving capacity in children and adolescents. *Zentralblatt für Didaktitk der Mathematik, 5*, 193-198.
Fischbein, E., Pampu, I. & Minzat, I.(1970). Effects of age and instruction on combinatorial ability in children. *The British Journal of Psychology, 40* (3), 261-270.
Fischbein, E., Pampu, I. & Minzat, I. (1975) The child's intuition of probability. Appendix II in Fischbein, E. *The intuitive sources of probabilistic thinking in children*. Dordrecht, The Netherlands: Reidel.
Fong, G. T., Krantz, D. H., & Nisbett, R. E. (1986). The effects of statistical training on thinking about everyday problems. *Cognitive Psychology, 18*, 253-292.
Frid, S. D. (1992). Calculus students' sources of conviction. In *Conference Proceedings Mathematics Education Research Group of Australasia* (pp. 294-304). Australia: MERGA.
Friel, S. N. & Bright, G. (1995). *Assessing students' understanding of graphs: Instruments and instructional module*. Chapel Hill, NC: UNC Mathematics and Science Education Network.
Friel, S. N. & Corwin R. B. (1990). Implementing the Standards: The statistics standards in K-8 mathematics. *Arithmetic Teacher, 38*(2), 35-39.
Friel, S. N. & Joyner, J. (1997). *Teach-Stat for teachers: Professional development manual*. Palo Alto, CA: Dale Seymour Publications.
Friel, S. N., Mokros, J. R., & Russell, S. J. (1992). *Statistics: Middles, means and in-betweens*. Palo Alto, CA: Dale Seymour Publications.
Friel, S. N., Russell, S., & Mokros, J. R. (1990). *Used Numbers: Statistics: middles, means, and in-betweens*. Palo Alto, CA: Dale Seymour Publications.
Frith, D. S., & MacIntosh, H. G. (1984). *A teacher's guide to assessment*. Cheltenham, UK: Stanley Thornes Publishers Ltd.
Gal, I. (1993). Reaching out: Some issues and dilemmas in expanding statistics education. In L. Pereira-Mendoza (Ed.), *Introducing data-analysis in the schools: Who should teach it and how?* (pp. 189-203). Voorburg, The Netherlands: International Statistics Institute.
Gal, I. (1994). *Assessment of interpretive skills*. Summary of Working Group at the Conference on Assessment Issues in Statistics Education, Philadelphia, Pennsylvania, September, 1994.
Gal, I. (1995). Statistical tools and statistical literacy: The case of the average. *Teaching Statistics, 17*(3), 97-99.
Gal, I. (in press). Assessing statistical knowledge as it relates to students' interpretation of data. In S. Lajoie (Ed.), *Reflections on statistics: Agendas for learning, teaching, and assessment in school contexts*. Hillsdale, NJ: Lawrence Erlbaum.
Gal, I., & Baron, J. (1996). Understanding repeated simple choices. *Thinking and Reasoning, 2*(1), 1-18.

Gal, I., & Ginsburg, L. (1994). The role of beliefs and attitudes in learning statistics: Toward an assessment framework. *Journal of Statistics Education* [Online], 2 (2).

Galbraith, P (1993). Paradigms, problems and assessment: some ideological implications. In M. Niss (Ed.), *Investigations into assessment in mathematics education: an ICMI study* (pp.73–86). Dordrecht, The Netherlands: Kluwer.

Garfield, J. B. (1991). Evaluating students understanding of statistics: Developing the Statistical Reasoning Assessment. In R. G. Underhill (Ed.), *Proceedings of the Thirteenth annual meeting of the Psychology in Mathematics Education group*. Vol. 2.

Garfield, J. B. (1993). An authentic assessment of students' statistical knowledge. In N. L. Webb & A. F. Coxford (Eds.), *Assessment in the mathematics classroom* (pp. 187-196). Reston, VA: National Council of Teachers of Mathematics.

Garfield, J. (1994). Beyond testing and grading: using assesment to improve students's learning. *Journal of Statistics Education*, [Online]2 (1).

Garfield, J. (1995a). How students learn statistics. *International Statistical Review*. 63 (1), 25-34.

Garfield, J. (1995b). Key ideas of Randomness and Uncertainty. Keynote speech at SciMath writing conference. Chaska, MN.

Garfield, J. (1996). Assessing student learning in the context of evaluating a Chance Course. *Communications in Statistics: Theory and Methods*, 25 (11), 2863-2873.

Garfield, J., & Ahlgren, A. (1988). Difficulties in learning basic concepts in probability and statistics: Implications for research. *Journal for Research in Mathematics Education, 19*, 44-63.

Garofalo, J. & Lester, F. K. Jr. (1985). Metacognition, cognitive monitoring, and mathematical performance. *Journal for Research in Mathematics Education, 16*, 163-176.

Gigerenzer, G., Swijtink, Z., Daston, L., Beatty, J., & Kruger, L. (1989). *The empire of chance: How probability changed science and everyday life*. Cambridge: Cambridge University Press.

Goodlad, J. (1984). *A place called school*. New York: McGraw Hill.

Graham, A. (1987). *Statistical investigations in the secondary school*. Cambridge: Cambridge University Press.

Graham, A. (1990). Statistical investigations—data handling to a purpose. *Teaching Statistics,12*, 2, 58-60.

Gravemeijer, K. (1994). Educational development and developmental research in mathematics education. *Journal for Research in Mathematics Education, 25*, 443-471.

Green, D. R. (1981). *Probability concepts in school pupils aged 11-16 years*. Unpublished dissertation. Loughborough University, England.

Green, D. R. (1983). A survey of probability concepts in 3000 students aged 11-16 years. In D. R. Grey, P. Holmes, V. Barnett, & G. M. Constable (Eds.), *Proceedings of the First International Conference on Teaching Statistics* (pp. 766-783). Sheffield, UK: Teaching Statistics Trust.

Green, K. E. (1993, April). Affective, evaluative, and behavioral components of attitudes toward statistics. Paper presented at the annual meeting of the American Educational Research Association, Atlanta, GA.

Green, K. E. (1994, April). The affective component of attitude in statistics instruction. Paper presented at the annual meeting of the American Educational Research Association, New Orleans, LA.

Greer, B., & Semrau, G. (1984). Investigating psychology students' conceptual problems in relation to learning statistics. *Bulletin of the British Psychological Society, 37*, 123-5.

Grimaldi, R. (1989). *Discrete and combinatorial mathematics. An applied introduction*. Reading, MA: Addison-Wesley.

Gronlund, N. E. (1993). *How to make achievement tests and assessments* (5th Edition). New York: Allyn and Bacon.
Guthrie, J. T., Weber, S., & Kimberly, N. (1993). Searching Documents: Cognitive processes and deficits in understanding graphs, tables, and illustrations. *Contemporary Educational Psychology, 18*, 186-221.
Hacking, I. (1990). *The taming of chance.* Cambridge, England: Cambridge University Press.
Hancock, C., Kaput, J. J., & Goldsmith, L. T. (1992). Authentic inquiry with data: Critical barriers to classroom implementation. *Educational Psychologist, 27*, 337-364.
Harnisch, D. L., Sato, T., Zheng, P., Yamagi, S., and Connell, M. (1994, April). *Concept mapping approach and its applications in instruction and assessment.* Paper presented at the meeting of the American Educational Research Association, New Orleans, LA.
Hart, E. W. (1992). Discrete mathematics: An exciting and necessary addition to the secondary school curriculum. In M. J. Kenney and C. R. Hirsch (Eds.), *Discrete mathematics across the curriculum, K-12* (pp. 67-77), Reston, VA: National Council of Teachers of Mathematics.
Hawkins, A. (1996). Myth-Conceptions. In C. Batanero (Ed.), *Proceedings for the International Association for Statistical Education roundtable on research on the role of technology in teaching and learning statistics* (pp. 11-24). Spain: University of Granada.
Hawkins, A., Jolliffe, F. & Glickman, L. (1992). *Teaching statistical concepts.* London: Longman.
Hawkins, A. S. & Kapadia, R. (1984). Children's conceptions of probability: A psychological and pedagogical review. *Educational Studies in Mathematics.* 15, 349-377.
Heitele, D. (1975). An epistemological view on fundamental stochastic ideas. *Educational Studies in Mathematics, 6,* 187-205.
Helgeson, S. L. (1993). *Assessment of science teaching and learning outcomes.* Monograph #6, March 1993. National Center for Science Teaching and Learning, Ohio State University.
Herman, J. L., Aschbacher, P. R., & Winters, L. (1992). *A practical guide to alternative assessment.* Alexandria, VA: Association for Supervision and Curriculum Development.
Hibbard, M. (1992). Bringing authentic performance assessment to life with cooperative learning. *Cooperative Learning, 13,* 30-32.
Hillyer, J. (1979). Statistics at CEE. *Teaching Statistics 1*(2), 56-59.
Holley, C. D., & Dansereau, D. F. (1984). *Spatial learning strategies: Techniques, applications, and related issues.* Orlando: Academic Press.
Horton, P. B., McConney, A. A., Gallo, M., Woods, A. L., Senn, G. J., & Hamelin, D. (1993). An investigation of the effectiveness of concept mapping as an instructional tool. *Science Education, 77,* 95-111.
Inhelder, B. & Piaget, J. (1955). *De la logique de l'enfant à la logique de l'adolescent.* [From the logic of child to the logic of adolescent]. Paris: Presses Universitaires de France.
Inspiration [Computer software]. (1994). Portland, OR: Inspiration Software, Inc.
Johnson, D. W., Johnson, R. T., & Smith, K. A. (1991). *Active learning: Cooperation in the college classroom.* Edina, MN: Interaction Book.
Johnson, P. J., Goldsmith, T. E., & Teague, K. W. (1995). Similarity, structure, and knowledge: A representational approach to assessment. In P. D. Nichols, S. F. Chipman, & R. L. Brennan (Eds.), *Cognitively diagnostic assessment* (pp. 221-249). Hillsdale, NJ: Lawrence Erlbaum.
Jolliffe, F. (1991). New approaches to statistics. In S. McAllister, & M. Speller (Eds.) *Proceedings, Mathematics in a Changing Culture Conference,* (pp. 1-8). Glasgow College, UK.

Jolliffe, F. (1993). The preliminary stages of data analysis at the school level. In Introducing data analysis in the schools: who should teach it and how? In L. Pereira-Mendoza (Ed.), *Introducing data-analysis in the schools: Who should teach it and how?* (pp. 87-96). Voorburg, The Netherlands: International Statistics Institute.

Jonassen, D. H., Beissner, K., & Yacci, M. (1993). *Structural knowledge: Techniques for representing, conveying, and acquiring structural knowledge.* Hillsdale, NJ: Lawrence Erlbaum.

Kader, G. (1992). Implementing the NCTM Standards in probability and statistics in grades 9-12: Project STAT-LINC. Paper presented at the 1992 Winter Meeting of the American Statistical Association, Louisville, KT.

Kader, G. & Perry, M. (1994). Learning statistics. *Mathematics teaching in the middle school, 1*(2), 130 - 136.

Kahneman, D. (1991). Judgment and decision making: A personal view. *Psychological Science, 2*(3), 142-145.

Kahneman, D. Slovic, P. & Tversky, A. (1982). *Judgement under uncertainty: Heuristics and biases.* Cambridge, England: Cambridge University Press.

Kahneman, D. & Tversky, A. (1982). Variants of uncertainty. *Cognition, 11*, 143-157.

Kapur, J. N. (1970). Combinatorial analysis and school mathematics. *Educational Studies in Mathematics, 3*, 111-127.

Kaput, J. J. (1992). Technology and mathematics education. In D. A. Grouws (Ed.), *Handbook for Research on Mathematics Teaching and Learning* (pp. 515-556). New York: Macmillan.

Kelly, A.V. (1978). *Mixed-ability grouping: Theory and practice.* London, England: Harper & Row.

Kimmel, M. L. (1992). Learning from industry: Using quality control techniques to monitor and motivate student progress. *The Statistics Teacher Network, 31*, 1-4.

Knight, P. (1992). How I use portfolios in mathematics. *Educational Leadership, 49*(8), 71-72.

Konold, C. (1990). *ChancePlus: A computer based curriculum for probability and statistics.* Annual report to NSF. Scientific Reasoning Research Institute, University of Massachusetts, Amherst.

Konold, C. (1991a). Informal conceptions of probability. *Cognition and Instruction, 6* (1), 59-98.

Konold, C. (1991b). Understanding students' beliefs about probability. In E. von Glaserfeld (Ed.), *Radical constructivism in mathematicseducation* (pp. 139-156). Netherlands: Kluwer.

Konold, C. (1995). Issues in assessing conceptual understanding in probability and statistics. *Journal of Statistics Education* [Online], *3*(1).

Konold, C., Pollatsek, A., Well, A., Lohmeier, J., & Lipson, A. (1993). Inconsistencies in students' reasoning about probability. *Journal for Research in Mathematics Education, 24* (5), 392-414.

Kroll, D. L., Masingila, J. O., & Mau, S. T. (1992). Grading cooperative problem solving. *Mathematics Teacher, 85*, 619-627.

Lajoie, S. P. (1995). A framework for authentic assessment in mathematics. In T. A. Romberg (Ed.), *Reform in School mathematics and authentic assessment* (pp. 19-37). New York: SUNY Press.

Lajoie, S. P. (in preparation). Understanding of statistics. In E. Fennema & T. Romberg (Eds.), *Classrooms that promote understanding.*

Lajoie, S. P., & Derry, S. J. (Eds.). (1993). *Computers as cognitive tools.* Hillsdale, NJ: Lawrence Erlbaum.

Lajoie, S. P., Jacobs, V. R., & Lavigne, N. C. (1995). Empowering children in the use of statistics. *Journal of Mathematical Behavior, 14* (4), 401-425.

Lajoie, S. P., & Lesgold, A. (1992). Dynamic assessment of proficiency for solving procedural knowledge tasks. *Educational Psychologist, 27* (3), 365-384.

Landwehr, J. M., Swift, J., & Watkins, A. E. (1987). *Exploring surveys and information from samples.* Palo Alto, CA: Dale Seymour Pub.

Landwehr, J. M., & Watkins, A. E. (1987). *Exploring data.* Palo Alto, CA: Dale Seymour publications.

Lane, S. (1993). The conceptual framework for the development of a mathematics performance assessment instrument. *Educational Measurement: Issues and Practice, 12*(2), 16-23.

Lappan, G., Fey J., Fitzgerald W., Friel S., & Phillips E. (1996). *Data about us.* (Connected Mathematics Project). Palo Alto, CA: Dale Seymour Publications.

Lavigne, N. C. (1994). *Authentic assessment: A library of exemplars for enhancing statistics performance.* Unpublished Master's Thesis. McGill University, Montreal, Quebec, Canada.

Lavigne, N. C., & Lajoie, S. P. (1996). Communicating performance standards to students through technology. *The Mathematics Teacher, 89* (1), 66-69.

Leder, G. (1992). Assessment: as we sow we reap. In M.Stephens & J. Izard, J. (Eds.), *Reshaping assessment practices: Assessment in the mathematical sciences under challenge* (pp. 114–124). Hawthorn VIC: Australian Council for Educational Research. (Proceedings from the First National Conference on Assessment in the Mathematical Sciences, Geelong, November 1991).

Lehrer, R., & Romberg, T. (1996). Exploring children's data modeling. *Cognition and Instruction, 14* (1), 69-108.

Lesh, R., Hoover, M., & Kelly, A.E. (1992). Equity, Assessment, and Thinking Mathematically: Principles for the Design of Model-Eliciting Activities. In I. Wirszup & R. Streit (Eds.), *Proceedings of the UCSMP International Conference on Mathematics Education: Developments in School Mathematics Education Around the World: Volume 3* (pp. 104-129). Reston, VA: National Council of Teachers of Mathematics.

Lesh, R., & Lamon, S. J. (1992) (Eds.) *Assessment of authentic performance in school mathematics.* Washington, DC: American Association for the Advancement of Science.

Linn, R. L., Baker, E. L., & Dunbar, S. B. (1991). Complex, performance-based assessment: Expectations and validation criteria. *Educational Researcher, 20* (8), 15-21.

Lovie, P. (1978). Teaching intuitive statistics II: Aiding the estimation of standard deviations. *International Journal of Mathematical Education in Science and Technology, 9,* 213-219.

Lovie, P., & Lovie, A. D. (1976). Teaching intuitive statistics I: Estimating means and variances. *International Journal of Mathematical Education in Science and Technology, 7,* 29-40.

Macdonald-Ross, M. (1977). How numbers are shown. *AV Communication Review, 25,* 359-409.

Margolin, B. H. (1988). Statistical aspects of using biologic markers. *Statistical Science, 3,* 351-357.

Marshall, S. (1995). *Schemas in problem solving.* Cambridge: Cambridge University Press.

Mathematical Sciences Education Board (1993). *Measuring up: Prototypes for mathematics assessment.* Washington, DC: National Academy Press.

Mathematical Sciences Education Board (MSEB) (1993). *Measuring what counts: A policy brief.* Washington, DC: National Academy Press.

McGregor, M. (1993). Mathematical report writing - what value? *Australian Mathematics Teacher, 49*(1), 31.

McKnight, C. C. (1990). Critical evaluation of quantitative arguments. In G. Kulm (Ed.), *Assessing higher order thinking in mathematics* (pp. 169-185). Washington, DC: American Association for the Advancement of Science.

McLeod, D. B. (1992). Research on affect in mathematics education: A reconceptualization. In D. A. Grouws (Ed.), *Handbook of Research on Mathematics Teaching and Learning.* (pp. 575-596). NY: Macmillan.

Meece, J. L., Wigfield, A., & Eccles, J. S. (1990). Predictors of math anxiety and its influence on young adolescents' course enrollment intentions and performance in mathematics. *Journal of Educational Psychology, 82*, 60-70.

Millar, M. G., & Millar, K. U. (1990). Attitude change as a function of attitude type and argument type. *Journal of Personality and Social Psychology, 59*, 217-228.

Miller, E. (1992). Tips on assessment and evaluation for cooperative work. *Cooperative Learning, 13*, 48-49.

Ministry of Education. (1992). *The New Zealand curriculum framework.* Wellington: Ministry of Education.

Ministry of Education. (1993). *Mathematics in the New Zealand curriculum.* Wellington: Ministry of Education.

Montgomery, H., & Allwood, C. M. (1978a). On the subjective representation of statistical problems. *Scandinavian Journal of Educational Research, 22*, 102-127.

Montgomery, H., & Allwood, C. M. (1978b). Subjective confidence in the correctness of statistical problem solutions. *Goteborg Psychological Reports, 8*(1).

Moore, D. S. (1990). Uncertainty. In L. A. Steen (Ed.), *On the shoulders of giants: New approaches to numeracy* (pp. 95-137). Washington, DC: National Academy Press.

Moore, D.S.(1992). Teaching statistics as a respectable subject. In F. & S. Gordon (Eds.), *Statistics for the twenty-first Century.* (pp. 14-25). Washington, DC: The Mathematical Association of America.

Moore, D.S. (in press). New pedagogy and new content: The case of statistics. *International Statistical Review.*

Morris, C. (1989).*Quantitative approaches in business studies, 2nd edition.* London: Pitman Publishers.

National Council of Teachers of Mathematics. (1989). *Curriculum and evaluation standards for school mathematics.* Reston, VA: Author.

National Council of Teachers of Mathematics. (1991). *Professional standards for teaching mathematics.*Reston, VA: Author.

National Council for Teachers of Mathematics. (1995). *Assessment standards for school mathematics.* Reston, VA: Author.

Naveh-Benjamin, M., Lin, Y., & McKeachie, W. J. (1995). Inferring students' cognitive structures and their development using the "Fill-in-the-Structure" (FITS) technique. In P. D. Nichols, S. F. Chipman, & R. L. Brennan (Eds.), *Cognitively diagnostic assessment* (pp. 279-304). Hillsdale, NJ: Lawrence Erlbaum.

Nickerson, R. S. (1995). Can technology help teach for understanding? In D. N. Perkins, J. L. Schwartz, M. M. West, M. S. Wiske (Eds.), *Software goes to school: Teaching for understanding with new technologies* (pp. 7-22). NY: Oxford University Press.

Nisbett, R.E., Krantz, D. H., Jepson, C. & Kunda, Z. (1983). The use of statistical heuristics in everyday inductive reasoning. *Psychological Review. 90* (4), 339-363.

Nitko, A. J., & Lane, S. (1991). Solving problems is not enough: Assessing and diagnosing the ways in which students organise statistical concepts. In D.Vere-Jones (Ed.), *Proceedings of the Third International Conference on Teaching Statistics, Volume 1: School and General Issues* (pp. 467-474). Voorburg, The Netherlands: International Statistical Institute.

Northern Examinations and Assessment Board (1994). *Information technology syllabus.* Manchester, England: Author.

Novak, J. D., & Gowin, D. B. (1984). *Learning how to learn.* New York: Cambridge University Press.

Novak, J. D., & Musonda, D. (1991). A twelve year longitudinal study of science concept learning. *American Educational Research Journal, 28*, 117-153.

Nunez, R. (1993). Approaching infinity: A view from cognitive psychology. *Proceedings of the Fifteenth Annual Meeting of the North American Chapter of the International Group for the Psychology of Mathematics Education., Volume 1*, 105-111.

Ohio Math Project (1992). *Introduction to algebra and statistics.* Dayton, OH: EPA Associates.

Pandey, T. (1990). Power items and the alignment of curriculum and assessment. In G. Kulm (Ed.), *Assessing higher order thinking in mathematics* (pp. 39-51).Washington, DC: American Association for the Advancement of Science.

Paulu, N. (1994). *Improving math and science assessment: Report on the secretary's third conference on mathematics and science education.* Washington, DC: US Department of Education.

Pereira-Mendoza, L. & Mellor, J. (1991). Students' concepts of bar graphs: Some preliminary findings. In D. Vere-Jones (Ed.), *Proceedings of the Third International Conference on Teaching Statistics. Volume 1: School and general issues* (pp. 150-157). Voorburg, The Netherlands: International Statistical Institute.

Perkins, D. N., Crismond, D., Simmons, R., & Unger, C. (1995). Inside Understanding. In D. N. Perkins, J. L.Schwartz, M. M. West, M. S. Wiske (Eds.), *Software goes to school: teaching for understanding with new technologies* (pp.70-87). New York: Oxford University Press.

Pesci, A. (1994). Tree graphs: visual aids in causal compound experiments. In J. P.Ponte & J. F. Matos (Eds.). *Proceedings of the Psychology of Mathematics Education Annual Meeting, Volume 4* (pp. 25-32). Lisbon: University of Lisbon.

Phillips, B., & Jones, P., (1991). Devloping statistical concepts for engineering students using computer packages. In D. Vere-Jones (Ed.), *Proceedings of the Third International Conference on Teaching Statistics, Volume 2: Teaching statistics beyond school level* (pp. 255-260). Voorburg, The Netherlands: International Statistical Institute.

Piaget, J. (1987). *Possibility and Necessity.* Vol. 2: The role of necessity in cognitive development. (Originally published in French in 1983) Translated from the French by H. Feider.

Piaget , J. & Inhelder, B. (1951). *La génese de l'idée d' hasard chez l'enfant.* [The origin of the idea of chance in children]. París: Presses Universitaire de France.

Piaget, J. , & Inhelder, B. (1975). *The origin of the idea of chance in children.* London: Routledge & Kegan Paul.

Pinker, S. (1990). A theory of graph comprehension. In R. Freedle (Ed.), *Artificial intelligence and the future of testing* (pp. 73-126). Hillsdale, NJ: Lawrence Erlbaum.

Pollatsek, A., Lima, S., & Well, A. D. (1981). Concept or computation: Students' understanding of the mean. *Educational Studies in Mathematics, 12*, 191-204.

Polya, G. (1945). *How to solve it.* Garden City, NY: Doubleday.

Popham, W. J. (1990). *Modern measurement: A practitioner's guide.* London: Prentice Hall.

Pretorius, T. B., & Norman, A. M. (1992). Psychometric data on the Statistics Anxiety Scale for a sample of South African students. *Educational and Psychological Measurement, 52*, 933-937.

Pulaski, M. A. (1980). *Understanding Piaget.* New York: Harper & Row.

Rachlin, S. (1992). Hawaii Algebra: A curriculum research and development model. In T. Cooper (Ed.), *From numeracy to algebra: The proceedings of the second mathematics teaching and learning conference,* (pp. 34-49). Brisbane, Australia: QUT Centre for Mathematics and Science Education.

Resnick, L. B. (1987). *Education and learning to think.* Washington, DC: National Academy Press.

Resnick, L. B. (1988). Treating mathematics as an ill-structured discipline. In R. Charles & E. Silver (Eds.), *The teaching and assessing of mathematical problem solving: Multiple research perspectives,* (pp. 32-60). Reston, VA: National Council of Teachers of Mathematics.

Resnick, L. B. & Klopfer, L. E. (1989). Toward the thinking curriculum: An overview. In L. B. Resnick & L. E. Klopfer (Eds.), *Toward the thinking curriculum: Current cognitive research* (1989 Yearbook of the Association for Supervision and Curriculum Development). Alexandria, VA: ASCD.

Roberts, D. M., & Reese, C. M. (1987). A comparison of two scales measuring attitudes towards statistics. *Educational and Psychological Measurement, 47*, 759-764.

Roberts, D. M., & Saxe, J. E. (1982). Validity of a statistics attitude survey: A follow-up study. *Educational and Psychological Measurement, 42*, 907-912.

Romberg, T. (1989). Evaluation: A coat of many colors. In D. Robitaille (Ed.), *Evaluation and assessment in mathematics education* (pp. 3-17). Paris: UNESCO.

Romberg, T. A. (1993). How one comes to know: models and theories of the learning of mathematics. In M. Niss (Ed.,) *Investigations into assessment in mathematics education: An ICMI study* (pp. 97–111). Dordrecht, The Netherlands: Kluwer.

Romberg, T. A. Allison, J., Clarke, B., Clarke, D., & Spence, M. (Nov., 1991). *School Mathematics Expectations: A comparison of curricular documents of eight countries with the NCTM Standards of the US* Paper prepared for the New Standards Project.

Romberg, T. A., Zarinnia, E. A. & Collis, K. F. (1991). A new world view of assessment in mathematics. In G. Kulm (Ed.), *Assessing higher rder thinking in mathematics* (pp. 21-38). Washington: American Association for the Advancements of Sciences.

Rosebery, A. S., & Rubin, A. (1989). Reasoning under uncertainty: Developing statistical reasoning. *Journal of Mathematical Behavior, 8*, 205-219.

Rubin, A., Bruce, B., & Tenney, Y. (1991). Learning about sampling: Trouble at the core of statistics. In D. Vere-Jones (Ed.), *Proceedings of the Third International Conference on Teaching Statistics. Volume 1: School and general issues* (pp. 314-319). Voorburg: International Statistical Institute.

Rubin, A. & Rosebery, A. (1990). Teacher's misunderstandings in statistical reasoning: Evidence from a field test of innovative materials. In A. Hawkins (Ed.) *Training teachers to teach statistics.* Voorburg, The Netherlands: International Statistical Institute.

Rubin, A., Rosebery, A. S., & Bruce, B. (1988). *ELASTIC and reasoning under uncertainty: Final report.* Cambridge, MA: Bolt Beranek & Newman.

Ruiz-Primo, M. A., & Shavelso, R. J. (1996). Problems and issues in the use of concept maps in science assessment. *Journal of Research in Science Teaching, 33*, 569-600.

Russell, S. J., & Corwin, R. (1989a). *Used Numbers: Real data in the classroom.* Palo Alto, CA: Dale Seymour Publications.

Russell, S. J. & Corwin, R. (1989b). *Statistics: The shape of the data.* Palo Alto, CA: Dale Seymour Publications.

Salomon, G., Perkins, D. N., & Globerson, T. (1991). Partners in cognition: Extending human intelligence with intelligent technologies. *Educational Researcher, 20*, 10-16.

SCANS (1992). *Learning a living: a blueprint for high performance.* Secretary's Commission on Achieving Necessary Skills, US Department of Labor. Washington, DC: US Government Printing Office.

Schau, C., Dauphinee, T., & Del Vecchio, A. (1992, April). The development of the Survey of Attitudes Toward Statistics. Paper presented at the annual meeting of the American Educational Research Association, San Francisco, CA.

Schau, C., Stevens, J., Dauphinee, T. L., & Del Vecchio, A. (1995). The development and validation of the Survey of Attitudes Toward Statistics. *Educational and Psychological Measurement, 55*, 868-875.

Scheaffer, R. L. (1988). Statistics in the schools: The past, present and future of the quantitative literacy project. *Proceedings of the American Statistical Association from the Section on Statistical Education* (pp. 71-78). Washington, DC: American Statistical Association.

Schifter, D. & Fosnot, C. (1992). *Reconstructing mathematics education: Stories of teachers meeting the challenge of reform.* New York: Teachers College Press.

Schifter, D. & Simon, M. (1992). Assessing teachers' development of a constructivist view of mathematics learning. *Teaching & Teacher Education, 8*(2), 187-197.

Schoenfeld. A. H. (1985). *Mathematical problem solving.* Orlando, FL: Academic Press.

Schoenfeld, A. H. (1987). What's all the fuss about metacognition? In A. H. Schoenfeld (Ed.), *Cognitive science and mathematics education* (pp. 189- 215). Hillsdale, NJ: Lawrence Erlbaum.

Schoenfeld, A. H. (1992). Learning to think mathematically: Problem solving, metacognition, and sense making in mathematics. In D. A. Grouws (Ed.), *Handbook of Research on Mathematics Teaching and Learning* (pp. 334-370). New York: Macmillan.

Schwartz, J. L. (1989). Intellectual mirrors; A step in the direction of making schools into knowledge-making places. *Harvard Educational Review, 59* (1), 51-61.

Shaughnessy, J. M. (1992). Research on probability and statistics: Reflections and directions. In D.A. Grouws (Ed.) *Handbook of research on mathematics teaching and learning*, (pp. 465-494). New York: Macmillan.

Shavelson, R. J. & Baxter, G. (1992). What we've learned about assessing hands-on science. *Educational Leadership, 49*(8), 20-25.

Silver, E. A. (1987). Foundations of cognitive theory and research for mathematics problem solving instruction. In A. H. Schoenfeld (Ed.), *Cognitive science and mathematics education* (pp. 33-60). Hillsdale, NJ: Lawrence Erlbaum.

Skemp, R. R. (1979). *Intelligence, learning and action.* Chichester, UK: Wiley.

Skemp, R. R. (1987). *The psychology of learning mathematics.* Hillsdale, NJ: Lawrence Erlbaum.

Smith, J. T. & Griffin, M. P. (1991). The use of video feedback in training statistical consultants. In D. Vere-Jones (Ed.), *Proceedings of the Third International Conference on Teaching Statistics, Volume 2* (pp. 461-468). Voorburg, The Netherlands: International Statistical Institute.

Smith, J., diSessa, A., & Roschelle, J. (1993). Misconceptions reconceived: A constructivist analysis of knowledge in transition. *Journal of the Learning Sciences,* 3(2), 115-163.

Stenmark, J. K. (1989). *Assessment alternatives in mathematics: an overview of assessment techniques that promote learning.* Berkeley CA: EQUALS and the Californian Mathematics Council Campaign for Mathematics.

Stenmark, J. K. (1991). *Mathematics assessment: Myths, models good questions and practical suggestions.* Reston VA: National Council of Teachers of Mathematics.

Stiggins, R. J. (1987). Design and development of performance assessments. *Educational Measurement: Issues and Practice,* 6 (3), 33-42.

Stodolsky, S. S. (1985). Telling math: Origins of math aversion and anxiety. *Educational Psychologist.* 20 (3), 123-133.

Stone, A., & Russell, S. J. (1992).*Counting ourselves and our family.* Palo Alto, CA: Dale Seymour Publications.

Surber, J. R. (1984). Mapping as a testing and diagnostic device. In C. D. Holley & D. F. Dansereau (Eds.), *Spatial learning strategies: Techniques, applications, and related issues* (pp. 213-233). Orlando: Academic Press.

Swets, J. A., Rubin, A., & Feurzeig, W. (1987) *Cognition, computers, and statistics: software tools for curriculum design, Report No. 6447,* Cambridge, MA: BBN, Inc.

That's Life. (1993, July 21). *Hobart Mercury,* p. 17.

Tobias, S. (1993). *Overcoming math anxiety.* New York: Norton.

Travers, K., Stout, W., Swift, J., & Sextro, J. (1985). *Using statistics.* Menlo-Park, CA: Addison-Wesley.

Tufte, E. R. (1983). *The visual display of quantitative information.* Chesire, CT: Graphics Press.

Tversky, A. & Kahneman, D. (1971). Belief in the law of small numbers. *Psychological Bulletin,* 2, 105-110.

Tversky, A. & Kahneman, D. (1974). Judgment under uncertainty: Heuristics and biases. *Science, 185,* 1124-1131.

Vallone, R. & Tversky, A. (1985). The hot hand in basketball: On the misperception of random sequences. *Psychological Review,* 90, 293-315.

Vermont Department of Education (1991). *Looking beyond "the answer": The report of Vermont's mathematics portfolio assessment program.* Montpelier, Vermont.

von Glasersfeld, E. (1987). Learning as a constructive activity. In C. Janvier (Ed.), *Problems of representation in the teaching and learning of mathematics* (pp. 3-17). Hillsdale, NJ: Lawrence Erlbaum.

von Glasersfeld, E. (1989). Constructivism in education. In T. Husén, T. N. Postlewaite (Eds.), *The international encyclopaedia of education: Research and studies.* Supplementary Volume (pp. 162–163). New York: Pergamon Press.

Wainer, H. (1980). A test of graphicacy in children. *Educational Researcher.* 21(1), 14-23.

Walklin, L. (1991). *Instructional Techniques & Practice.* Cheltenham, UK: Stanley Thornes Publishers Ltd.

Wallman, K.K. (1993). Enhancing statistical literacy: Enriching our society. *Journal of the American Statistical Association,* 88, 1-8.

Waters, L. K., Martelli, A., Zakrajsek, T., & Popovich, P. M. (1988). Attitudes toward statistics: An evaluation of multiple measures. *Educational and Psychological Measurement,* 48, 513-516.

Watson, J. M. (1992). Fishy statistics. *Teaching Statistics,* 14 (3), 17-21.

Watson, J. M. (1993). Introducing the language of probability through the media. In M. Stephens, A. Wayward, D. Clarke, & J. Izard (Eds.), *Communicating mathematics – Perspectives from current research and classroom practice in Australia* (pp. 119-139). Melbourne: Australian Council for Educational Research.

Watson, J. M., Collis, K. F., Callingham, R. A., & Moritz, J. B. (In press). A model for assessing higher order thinking in Statistics. *Educational Research and Evaluation.*

Webb, N. L. (1992). Assessment of students' knowledge of mathematics: Steps toward a theory. In D. A. Grouws (Ed.), *Handbook of research on mathematics teaching and learning* (pp. 661-683). New York: Macmillan.

Webb, N. L. (1993). Assessment for the mathematics classroom. In N. L. Webb & A. F. Coxford (Eds.), *Assessment in the Mathematics Classroom, 1993 Yearbook* (pp. 1-6). Reston, VA: National Council of Teachers of Mathematics.

Webb, R., O'Meara, M, & Brown, B. (1993, January 28). Coles Myer accelerates retail purge. *Australian Financial Review*, p. 1.

Weight guidelines too lenient: study. (1995, February 10). *Hobart Mercury*, p. 17.

Wild, C. J. (1994). Embracing the "wider view" of statistics. *The American Statistician, 48,* 163-171.

Wild, C. J. (1995). Continuous improvement of teaching: A case study in a large statistics course. *International Statistical Review, 63,* 49-68.

Wild, C. J., Triggs, C.M., & Pfannkuch, M. (1994). Assessment on a budget: Using traditional methods imaginitively. Technical Report STAT9411 (Revised 1995), Department of Statistics, University of Auckland.

Wiliam, D. (1994). Assessing authentic tasks: alternatives to mark schemes. *Nordic Studies in Mathematics Education* 2(1), 48–68.

Wise, S. L. (1985). The development and validation of a scale measuring attitudes towards statistics. *Educational and Psychological Measurement, 45,* 401-405.

Woehlke, P. L. (1991, April). An examination of the factor structure of Wise's Attitudes Toward Statistics scale. Paper presented at the annual meeting of the American Educational Research Association, Chicago, IL.

Zessoules, R. & Gardner, H. (1991). Authentic assessment: Beyond the buzzword and into the classroom. In V. Perrone (Ed.), *Expanding student assessment.* Alexandria, VA: Association for Supervision and Curriculum Development.

Author Index

Ahlgren, A.	1, 2, 91, 253	Clayden, A.D.	85
Allison, J.	223	Clayton, S.	16
Allwood, C.M.	85	Cobb, G.W.	1, 2, 3, 253, 254
Anderson, S.	165	Cockcroft, W.	23, 24
Armour-Thomas, E.	123, 126, 129	Cohen, S.	255, 256, 258, 260, 262
Arter, J.	69	Collins, A.	182
Artzt, A.F.	123, 126, 129, 136	Collis, K.F.	108, 219
Aschbacher, P.R.	28, 168, 169	Connell, M.	95
Baker, E.L.	126, 169, 186	Cook, R.	255
Ball, S.	165	Cooper, W.	176
Baron, J.	3	Corwin R.B.	50, 61, 117
Barron, B.	183	Crismond, D.	180
Batanero, C.	240, 245, 246, 247	Croft, M.R.	85
Baxter, G.	176	Curcio, F.R.	57, 113, 123, 124, 137
Beatty, J.	230	Dallal, G.E.	90
Begg, A.	20	Dansereau, D.F.	93
Beissner, K.	93	Daston, L.	230
Ben-Zvi, D.	182	Dauphinee, T.L.	42, 43, 44
Bentz, H.	192	Davies, J.	176
Biggs, J.B.	108	Davis, R.B.	167
Borovcnik, M.	192	Del Vecchio, A.M.	41, 42, 48
Brandt, R.	176	Derry, S.J.	180
Bransford, J.	169, 183	diSessa, A.	260
Bright, G.W.	57, 189	Dossey, J.A.	124
Brown, B.	113	Dubois, J.G.	245, 246
Brown, J.S.	182	Dunbar, S.B.	186
Bruce, B.	117	Eccles, J.S.	48
Burns, G.	255, 256, 258	Edwards, K.	40
Burrill, G.	3	Elliott, G.D.	145
Callingham, R.A.	108	Engel, A.	243
Carnevale, A.P.	108	English, F.	169
Carpenter, T.	180	Ewing, T.	109
Case, R.	108	Fischbein, E.	231, 235, 242, 245, 247, 249
Castles, I.	107	Fong, G.T.	85
Charles, R.	111, 120	Fosnot, C.	167
Chechile, R.	255, 256, 258	Freeman, M.	169
Clark, M.	203	Frid, S.D.	220
Clarke, B.	223	Friedlander, A.	182
Clarke, D.	223	Friel, S.N.	1, 51, 111, 123, 137, 189, 232

Frith, D.S.	209, 218	Jepson, C.	225, 234
Gal, I.	1, 3, 7, 41, 45, 110, 244	Johnson, D.W.	177
Galbraith, P.	20	Johnson, P.J.	103
Gallo, M.	99	Johnson, R.T.	177
Gardner, H.	176	Jolliffe, F.	194
Garfield, J.	1, 3, 4, 45, 91, 123, 200, 243, 251, 253	Jonassen, D.H.	93, 99
		Jones, P.	195
Garofalo, J.	126	Kader, G.	55
Gazit, A.	242, 245, 247, 249	Kahneman, D.	85, 228, 230
Gigerenzer, G.	230	Kapadia, R.	226
Ginsburg, L.	1, 41, 244	Kapur, J.N.	241
Globerson, T.	180	Kaput, J.J.	180, 181, 186
Goin, L.	183	Kelly, A.E.	66
Goldsmith, L.T.	181	Kelly, A.V.	140, 151
Goldsmith, T.E.	103	Kimberly, N.	124
Goodlad, J.	234	Kimmel, M.L.	203
Gowin, D.B.	93, 99	Klopfer, L.E.	169
Graham, A.	55	Knight, P.	176
Gravemeijer, K.	56	Konold, C.	1, 45, 85, 181
Green, D.R.	85, 243	Krantz, D.H.	85, 225, 234
Green, K.E.	40, 44, 47, 48	Kroll, D.L.	123
Greer, B.	85	Kruger, L.	230
Griffin, M.P.	202	Kulewicz, S.	183
Grimaldi, R.	246	Kunda, Z.	225, 234
Gronlund, N.E.	215	Lajoie, S.P.	179, 180, 181, 184
Guthrie, J.T.	124	Landwehr, J.M.	1, 117
Hacking, I.	230	Lane, S.	175, 192
Hamelin, D.	99	Lappan, G.	56
Hancock, C.	181	Lavigne, N.C.	181, 184
Harnisch, D.L.	95	Leder, G.	20
Hart, E.W.	239	Lehrer, R.	180, 181
Hasselbring, T.	183	Lesgold, A.	180
Hawkins, A.	180, 182, 192, 197, 226, 241, 247	Lesh, R.	66
		Lester, F.	111, 126
Heitele, D.	241	Lima, S.	87
Helgeson, S.L.	47	Lin, Y.	102
Herman, J.L.	28, 168, 169	Linn, R.L.	186
Hibbard, M.	123	Lipson, A.	181
Hillyer, J.	192	Littlefield, J.	183
Holley, C.D.	93	Lohmeier, J.	181
Hoover, M.	66	Lovie, A.D.	89
Horton, P.B.	99	Lovie, P.	89
Inhelder, B.	231, 241, 242	Macdonald-Ross, M.	137
Jacobs, V.R.	181	MacIntosh, H.G.	209, 218

Maher, C.A.	167	Pollatsek, A.	87, 181
Margolin, B.H.	210	Polya, G.	126
Marshall, S.	92, 93, 95	Popham, W.J.	215
Martelli, A.	48	Popovich, P.M.	48
McConney, A.A.	99	Pretorious, T.B.	44
McGregor, M.	112	Pulaski, M.A.	167
McKeachie, W.J.	102	Rachlin, S.	61
McKnight, C.C.	124	Reese, C.M.	43
McLeod, D.B.	40, 41, 47, 48	Resnick, L.B.	165, 169, 234
Meece, J.L.	48	Roberts, D.M.	43
Mellor, J.	113	Romberg, T.	181, 219, 223, 243, 247
Millar, K.U.	40	Roschelle, J.	260
Millar, M.G.	40	Rosebery, A.S.	181, 261
Miller, E.	123	Rubin, A.	117, 181, 261
Minzat, I.	231	Ruiz-Primo, M.A.	99, 103
Mokros, J.R.	1, 232	Russell, S.J.	56, 61, 232, 235
Montgomery, H.	85	Salomon, G.	180
Moore, D.S.	1, 4, 6, 38, 86, 224, 234, 254	Sato, T.	95
Moritz, J.B.	108	Schau, C.	42, 43, 48, 48
Morris, C.	199	Scheaffer, R.L.	181
Murphy, R.	165	Schifter, D.	167
Musonda, D.	99	Schoenfeld, A.H.	46, 109, 111, 126, 129
Naveh-Benjamin, M.	102	Schwartz, J.L.	180
Newman, S.E.	182	Semrau, G.	85
Nickerson, R.S.	181	Senn, G.J.	99
Nisbett, R.E.	85, 225, 234	Sextro, J.	1
Nitko, A.J.	192	Shaughnessy, J.M.	1, 253
Noddings, N.	167	Shavelson, R.J.	99, 103, 176
Norman, A.M.	44	Silver, E.A.	126
Novak, J.D.	43, 99	Simmons, R.	180
Nunez, R.	226	Simon, M.	167
O'Daffer, P.	111	Skemp, R.R.	85, 92, 93
O'Meara, M.	113	Slovic, P.	230
Pampu, I.	231	Smith, G.	255, 256, 258
Pandey, T.	217	Smith, J.	260
Paulu, N.	137	Smith, J.T.	202
Pereira-Mendoza, L.	113	Smith, K.A.	177
Perkins, D.N.	180	Spandel, V.	169
Perry, M.	55	Spence, M.	223
Pesci, A.	248	Starkings, S.A.	145
Pfannkuch, M.	206, 207, 210, 219	Stenmark, J.K.	25, 27
Phillips, B.	195	Stiggins, R.J.	186
Piaget, J.	226, 231, 241, 242	Stodolsky, S.S.	223, 234
Pinker, S.	124	Stone, A.	235

Stout, W.	1	Waters, L.K.	48
Surber, J.R.	99	Watkins, A.E.	1, 117
Swets, J.A.	197	Watson, J.M.	108, 112, 117
Swift, J.	1, 117	Webb, N.L.	27, 111, 112, 121, 245, 247, 251
Swijtink, Z.	230		
Teague, K.W.	103	Webb, R.	113
Tenney, Y.	117	Well, A.D.	87, 181
Tobias, S.	45	Wigfield, A.	48
Travers, K.	1	Wild, C.J.	206, 207, 209, 210, 211, 219
Triggs, C.M.	206, 207, 210, 219	Wiliam, D.	22
Tsai, F.	255, 256, 258	Winters, L.	28, 168, 169
Tufte, E.R.	124	Wise, S.L.	43, 48
Tversky, A.	85, 228, 239	Woehlke, P.L.	48
Unger, C.	180	Woods, A.L.	99
Vallone, R.	228	Yacci, M.	93
Von Glasersfeld, E.	20, 167	Yamagi, S.	95
Vye, N.	169	Zakrajsek, T.	48
Wainer, H.	56	Zarinnia, E.A.	219
Walklin, L.	143	Zessoules, R.	176
Wallman, K.K.	120	Zheng, P.	95

Subject Index

Assessment
  alignment with curriculum   169, 186
  alternative approaches   6-7, 27, 165, 177, 203
  authentic   179, 182-189, 205
  external   153-164
  formative   112-120, 191
  integrative   124-136
  models   27, 140
  oral   189
  principles   22-25, 32-33, 158-159, 206
  process, steps in   166
  purposes   22, 136, 169, 191
  role of   139
  staged   142-146
  standards   32
  summative   120, 191
  types   22
Attitudes   37-51
  instruments   49-51
  methods of assessing   43-47
Authentic assessment   27-36
  obstacles to   35-36
Averages -- *see Central tendency*

Beliefs   37-51

Calculators -- *see Technology*
Central tendency   183, 192, 193, 194, 197
Chance   223-238
Cognitive apprenticeship model   182, 188
Cognitive constructions   227-230, 237
Combinatorics   239-249
Communication   19, 218-219
  articulation   186
  language   110-111, 209, 217, 219
  listening   89-90
  skills   191
Computational ability   112, 191, 194, 202
Computers -- *see Technology*
Computer output   173, 195

Concept map	93-104
creation of	99, 103
completion task	100-103
Confidence intervals	214, 215
Connections, conceptual	9, 91-104, 255
Constructivism	20, 165-167
Correlation	85, 195
Curriculum, goals	2, 107, 154-155, 159, 192, 205, 206, 223, 242, 253
Data	
analysis	214, 215
describing	86
interpretation	109-110, 113, 184
real	207, 217, 219
Distribution	
frequency	193
probability	253-262
sampling	196, 257, 258, 266
Errors in students' work	161-162
Exemplars	171, 172, 181, 184, 185
External assessment	153-164
Factual knowledge	10, 191, 192, 194, 202
Fairness	218
Frameworks	
assessment	7, 192
curricular	17
graphical interpretation	124
group problem solving	123-137
knowledge	55-63
Goals -- *see Curriculum goals, Learning goals, Process goals*	
Grading	79, 88, 111, 120, 143-146, 148-149, 159-160, 191, 200-202, 216
Graphs	123-137
comprehension of	9, 189
constructing	87-88, 187
Graphs	
interpreting	70-83, 113-117, 210-213
knowledge	56-63
representations	181, 183, 184, 186, 258
Group work	9, 13, 121, 123-137, 140, 171, 173, 184, 191, 192, 200-203
Inference, statistical	93-95
Interpretation	
data	109-110, 113, 184
concepts	209, 213, 215, 219

Investigations
   process of   2-5, 39, 201
   statistical   55-56, 180, 188, 199, 219

Learning goals   2-5, 39, 108-111, 168
Literacy   12, 108, 120

Marks -- *see Grading*
Mathematical process   18
Media, statistics in   107, 109, 192
Misconceptions   197, 216, 259, 260
Modeling   8, 183, 184
   mental models   92-93
Model-eliciting activities   65-83
   principles   66
Multiple-choice formats   10, 95, 97, 191, 192, 196, 197, 202, 203, 205-220

Open-ended formats   10, 46, 111, 191, 192, 198-200, 202, 209, 259
Oral assessment   189, 191, 202, 203

p-value   214
Portfolio assessment   10, 165-178, 251
Probability   11, 195, 223-262
   distributions   196, 253-262
   test items   192, 196, 224, 223
   types of   226, 243
Problem solving   18, 38-39, 65-83, 92-93, 123-137, 168-169, 182, 187, 188, 239, 242, 243
   errors in   247-251
Process goals   38
Project work   9, 139-151, 153-164, 184, 191, 200, 251

Quality control   203
Questioning students   21
Questioning skills   110, 136, 188, 211

Randomness   223-226, 229-234, 237
Random experiment   241
Random variation   258
Reasoning   12, 18
   errors in   85-88
Regression   196

Sampling   110-120, 194, 198, 245, 247
Schemas   92
Scoring   199-200
   performance   186, 187
   rubrics   31, 170-171, 178, 198, 200, 204
Self-assessment   137, 174, 180, 185

Significance, statistical	211, 212
Statistical thinking	107-121, 182, 197, 207, 208, 211, 218, 253
Surveys	117-120
Technology	10, 179-189
Calculators	20
Computers	10, 20, 191, 192, 195, 196, 203, 253-262
Assessing computer skills	195, 202
Software environments	11-12, 261
Uncertainty -- *see Chance*	
Understanding, conceptual	8, 12, 85-90, 91-104, 223, 256, 260
Variability	181, 183, 194-197, 225
Variance, analysis of	88-89, 213
Verbal problems -- *see Word problems*	
Word problems	98, 244, 251

LaVergne, TN USA
26 August 2009
156110LV00005B/18/A